AF871433

Nanolubricants

Scrivener Publishing
100 Cummings Center, Suite 541J
Beverly, MA 01915-6106

Publishers at Scrivener
Martin Scrivener (martin@scrivenerpublishing.com)
Phillip Carmical (pcarmical@scrivenerpublishing.com)

Nanolubricants

Generation and Applications

Edited by
Mohd Yusuf
School of Life & Natural Sciences, The Glocal University, Saharanpur, India
Lalit Prasad
School of Basic Sciences, Galgotias University, Greater Noida, India
and
Shafat Ahmad Khan
Department of Chemistry, Galgotias University, Greater Noida, India

Scrivener Publishing

WILEY

This edition first published 2024 by John Wiley & Sons, Inc., 111 River Street, Hoboken, NJ 07030, USA and Scrivener Publishing LLC, 100 Cummings Center, Suite 541J, Beverly, MA 01915, USA
© 2024 Scrivener Publishing LLC
For more information about Scrivener publications please visit www.scrivenerpublishing.com.

All rights reserved. No part of this publication may be reproduced, stored in a retrieval system, or transmitted, in any form or by any means, electronic, mechanical, photocopying, recording, or otherwise, except as permitted by law. Advice on how to obtain permission to reuse material from this title is available at http://www.wiley.com/go/permissions.

Wiley Global Headquarters
111 River Street, Hoboken, NJ 07030, USA

For details of our global editorial offices, customer services, and more information about Wiley products visit us at www.wiley.com.

Limit of Liability/Disclaimer of Warranty
While the publisher and authors have used their best efforts in preparing this work, they make no representations or warranties with respect to the accuracy or completeness of the contents of this work and specifically disclaim all warranties, including without limitation any implied warranties of merchantability or fitness for a particular purpose. No warranty may be created or extended by sales representatives, written sales materials, or promotional statements for this work. The fact that an organization, website, or product is referred to in this work as a citation and/or potential source of further information does not mean that the publisher and authors endorse the information or services the organization, website, or product may provide or recommendations it may make. This work is sold with the understanding that the publisher is not engaged in rendering professional services. The advice and strategies contained herein may not be suitable for your situation. You should consult with a specialist where appropriate. Neither the publisher nor authors shall be liable for any loss of profit or any other commercial damages, including but not limited to special, incidental, consequential, or other damages. Further, readers should be aware that websites listed in this work may have changed or disappeared between when this work was written and when it is read.

Library of Congress Cataloging-in-Publication Data

ISBN 978-1-119-86510-0

Cover image: Pixabay.Com
Cover design by Russell Richardson

Set in size of 11pt and Minion Pro by Manila Typesetting Company, Makati, Philippines

Printed in the USA

10 9 8 7 6 5 4 3 2 1

Contents

Preface **xiii**

1 An Insight into Nanolubrication and Nanolubricants **1**
Deepshikha Singh, Wasim Khan and Mohd Yusuf
1.1 Introduction 1
1.2 Advantages of Nanolubricants 2
1.3 Preparation of Nanolubricants 3
1.3.1 Methods of Nanolubricant Preparation 4
1.3.2 Types of Nanolubricants Based on Additives' Characteristics 6
1.4 Lubrication Mechanism 8
1.5 Tribological and Thermophysical Properties of Nanolubricants 9
1.5.1 Tribological Properties 10
1.5.2 Thermophysical Properties 11
1.6 Conclusion and Future Directions 13
References 13

2 Nanolubrication Chemistry and Its Application **17**
Smrita Singh, Ashutosh Singh Chauhan and Lalit Prasad
2.1 Introduction 18
2.2 Nanolubrication and Its Requirements 19
2.3 Synthesis of Nanoparticles 21
2.3.1 Physical Method 21
2.3.2 Chemical Methods 22
2.3.3 Biological Methods 22
2.4 Preparation of Nanofluids/Nanolubricants 24
2.4.1 One-Step Method 24
2.4.2 Two-Step Method 25
2.4.2.1 Disadvantages of the Two-Step Method 25
2.4.3 Dispersion of Nanoparticles in Lubricating Oils 25

2.4.4 Interaction Forces in a Nanofluids/ Nanolubricant System 26
2.4.4.1 Van der Waals Forces 26
2.4.4.2 Electrostatic or Electric Double Layer Force (EDL) 27
2.4.4.3 DLVO Theory 28
2.4.4.4 Capillary Forces 28
2.5 Mechanism of Nanolubrication 28
2.6 Nanoparticle Properties Necessary for Nanolubrication 30
2.6.1 Nanolubricating Film Properties 31
2.6.2 Nanoparticles in Nanolubricants 32
2.7 Advantages of Nanolubricants 33
2.8 Nanoparticles Ability to Boost Grease Performance 34
2.9 Tribological Performance of Nanolubricants 38
2.9.1 Mechanical Properties of a Tribological System 38
2.9.2 Physicochemical Properties of the Lubricant 39
2.10 Nanolubricants and Base Oils 40
2.10.1 Nanolubricants 40
2.10.2 Base Oils 40
2.11 Various Types of Nanoparticles as Lubricant Additives 42
2.11.1 Metal Oxides 43
2.11.2 Metal Sulfides 44
2.11.3 Carbon-Based Nanoparticles 44
2.11.4 Nanocomposites 45
2.11.5 Rare Earth Compounds 46
2.12 Recent Advancement in Nanolubrication 46
2.13 Conclusion and Future Outlook 47
References 48

3 Characterization Techniques for Nanolubricants Using Different Approaches 61
Priyanka Chhabra and Akshara Johari
3.1 Introduction 62
3.2 Nanoparticles as an Additive to Nanolubricants 63
3.3 Application of Nanolubricants 66
3.4 Preparation of Nanolubricants 67
3.5 Characterization Factors of Nanolubricants 69
3.6 Characterization Techniques Used for Nanolubricants 70
3.6.1 Morphology and Topography Analysis 70
3.6.1.1 Dynamic Light Scattering (DLS): Particle Size Analysis 70

3.6.1.2 Electron Microscopy 72
3.6.1.3 X-Ray Diffraction 73
3.6.1.4 Atomic Force Microscopy 73
3.6.1.5 UV-Visible Spectroscopy 78
3.6.1.6 Fourier-Transform Infrared Spectroscopy (FTIR) 82
3.6.1.7 Raman Spectroscopy 84
3.7 Conclusion 85
References 85

4 Metal-Based Nanolubricants: Current and Future Perspectives 89
Deepak Gupta, Chandra Kumar, Aakash Mathur, Shruti Mishra, Anis Ahmad, Namrata Deka, Priyanki Kalita and Milan Singh
4.1 Introduction 90
4.2 Synthesis Mechanism of NPs 92
4.2.1 Top-Down Methods 92
4.2.1.1 Ball Milling 92
4.2.1.2 Electrospinning 92
4.2.1.3 Lithography 93
4.2.1.4 Sputter Deposition 94
4.2.1.5 Pulsed Laser Deposition 95
4.2.2 Bottom-Up Approaches 96
4.2.2.1 Chemical Vapor Deposition (CVD) 96
4.2.2.2 Hydrothermal/Solvothermal Methods 97
4.2.2.3 Sol–Gel Method 98
4.2.2.4 Co-Precipitation 98
4.3 NPs as Potential Candidate for Lubricant Additive 98
4.3.1 Nanometal-Based Lubricant Additives 99
4.3.2 Coefficient of Friction (COF) and Anti-Wear Properties of Nanolubricants 99
4.3.3 Lubrication Mechanisms 100
4.3.4 Rolling Effect or Ball-Bearing Effect 100
4.3.5 Protective Film Formation 101
4.3.6 Mending Effect 102
4.3.7 Polishing Effect 102
4.3.8 Surface Modified NP for Nanolubrication 102
4.4 Methods to Enhance Dispersion Stability of Nanolubricants 103
4.4.1 Physical Method 103
4.4.2 Use of Surfactant 104

4.4.3 Stability by Modification on Surface 104
4.4.4 Metal-Based Nanolubricants 105
4.4.5 Transition Metal Dichalcogenides (TMDCs)-Based Nanolubricants 115
4.6 Conclusion 124
References 124

5 Transition Metal-Based Catalysts for Preparing Biomass-Based Lubricating Oils 135
Binitendra Naath Mongal and Himanshu Arora
5.1 Introduction 135
5.2 Synthesis of Biolubricants 137
5.2.1 Esterification 137
5.2.2 Transesterification 137
5.2.3 Hydrogenation 137
5.2.4 Simultaneous Hydrogenation–Esterification 138
5.3 Catalysts for Biolubricant Synthesis 138
5.3.1 Catalysts for Esterification 138
5.3.2 Catalysts for Transesterification Reaction 140
5.4 Conclusions 144
References 145

6 Effect of Integration of Nanostructured Semimetals on Lubrication Performance of Non-Edible Vegetable Oil-Based Biolubricants 149
Umar Farooq and Farha Naaz
6.1 Introduction 150
6.2 Lubrication and Lubricating Materials 152
6.3 Inedible Vegetable Oils-Based Biolubricants 154
6.3.1 Resources 156
6.3.2 Properties 156
6.3.3 Merits and Demerits of Vegetable Oil-Based Lubricants 157
6.4 Nanoparticle Additives to Enhance Tribological Performance of Non-Edible Vegetable Oil Lubricants 159
6.4.1 Tribological Performance-Based Categorization of Nanoparticles 161
6.4.2 Effect of Nanoparticle Dispersion Stability, Shape, Size, Surface, Concentration, and Kind of Tribo-Test on the Tribological Performance 161
6.4.2.1 Dispersion Stability 161
6.4.2.2 Shape of Nanoparticles 161

6.4.2.3 Size of Nanoparticles 162
6.4.2.4 Surface Functionalization 162
6.4.2.5 Nanoparticles Concentration 162
6.4.2.6 Nature of Tribo-Testing 163
6.5 Tribological Mechanisms of Nanoparticles 163
6.5.1 Ball-Bearing Effect 163
6.5.2 Protective Film Formation 163
6.5.3 Mending Effect or Self-Healing Effect 163
6.5.4 Polishing Effect 163
6.5.5 Semimetal-Based Nano-Biolubricants 164
6.5.6 Boron-Based Nanoadditives in Non-Edible Vegetable Oils-Based Lubricants 164
6.6 Conclusion 167
References 168

7 Zinc Oxide Nanomaterials—Synthesis, Characterization, and Applications Focused on Lubricating Behavior 173
Monika Chauhan, Diwakar Chauhan, Ajay Kumar and Arvind Kumar Jain
7.1 Introduction 174
7.2 Preparations 176
7.2.1 Synthesis of ZnO by Pulsed Laser Ablation Technique 176
7.2.2 Synthesis of ZnO by Chemical Vapor Deposition Method 177
7.2.3 Synthesis of ZnO by Anodization Method 179
7.2.4 Synthesis of ZnO by Electrophoretic Deposition Process 180
7.2.5 Hydrothermal Process for the Synthesis of ZnO 180
7.2.6 Synthesis of ZnO by Electrochemical Deposition Method 181
7.2.7 Preparation of ZnO by Using the Sol–Gel Technique 182
7.2.8 Synthesis of ZnO by Thermolysis Method 184
7.2.9 Synthesis of ZnO by Combustion Method 186
7.2.10 Synthesis of ZnO by Ultrasonic Method 186
7.2.11 Microwave-Assisted Combustion Method to Synthesize Zinc Oxide 187
7.2.12 Synthesis of ZnO by Co-Precipitation Method 188
7.2.13 Synthesis of ZnO by Green Synthesis Method 189

7.3 Characterization 191
7.4 Applications 197
References 199

8 Improvement in the Properties of Biodegradable Nanolubricants 209
Sandip Paul Choudhury, Pushpendra Singh Shekhawat, Debanjan Bhattacharjee and Umesh K. Dwivedi
8.1 Introduction 209
8.1.1 Why Biodegradable Lubricants? 210
8.1.2 Vegetable Oil-Based Lubricants 210
8.1.3 Synthetic Lubricants 211
8.1.4 Properties and Synthesis of Nanolubricants 212
8.2 Nanoparticles for Lubricants 213
8.3 Types of Biodegradable Nanolubricants 217
8.3.1 Vegetable Oil as a Biodegradable Lubricant 218
8.3.2 Additives-Based Biodegradable Nanolubricants 219
8.3.3 Water-Based Nanolubricants 222
8.4 Conclusion and Outlook 223
References 224

9 Nanodimensional Metal-/Metal Oxide-Incorporated Vegetable Oil-Based Biodegradable Lubricants: Environmental Benefits, Progress, and Challenges 229
Pooja Sharma, Umar Farooq, Syed Salman Ali and Kaneez Fatima
9.1 Introduction 230
9.2 Concept of Lubrication and Characteristics of a Lubricant 232
9.2.1 Friction 232
9.2.2 Wear 233
9.2.3 Lubrication Regimes 234
9.2.4 Characteristics of a Lubricant 235
9.3 Vegetable Oil-Based Biolubricants 236
9.3.1 Limitations of Vegetable Oils (VOs) as Lubricants 238
9.3.1.1 Auto-Oxidation 238
9.3.1.2 Photo-Oxidation of VOs 240
9.3.1.3 Thermal Oxidation of Vegetable Oils 241
9.4 Nanolubricants 241
9.4.1 Mending Mechanism 242
9.4.2 Rolling/Ball-Bearing Mechanism 242

9.4.3 Formation of Protective Films 242
9.4.4 Polishing 243
9.4.5 Types of Nanoadditives 243
9.4.6 Vegetable Oil Metal/Metal Oxide-Based Nanolubricants 244
9.5 Challenges for Sustainable Bio-Nanolubrication 250
9.6 Conclusion 250
References 251

Index 257

Preface

Lubrication is the lifeblood of machinery and the key to its smooth operation. In the world of mechanics and engineering, the role of lubricants cannot be overstated. They are the unsung heroes that reduce friction between surfaces in contact, thus preventing excessive heat generation during motion. Beyond this primary function, lubricants find their application in diverse areas, including power transmission, foreign object transportation, and the regulation of surface temperature.

In recent times, our collective consciousness has shifted towards sustainable and environmentally friendly practices, prompting a transition from conventional lubricants to more efficient and eco-conscious alternatives. Among these emerging solutions, nanolubricants have emerged as formidable contenders, reshaping the landscape of lubrication technology. Their adoption not only promises enhanced performance but also carries the added benefit of environmental responsibility through biodegradability.

This book delves into the multifaceted realm of nanolubricants, exploring their characterization and application across various domains. From vegetable oil-based lubricants to those incorporating metal and non-metal oxide components, this comprehensive work encompasses nine meticulously curated chapters.

A particular focus is placed on the intriguing synergy between nano-dimensionality and the incorporation of metals and metal oxides into vegetable oil-based biodegradable lubricants. The book explores the environmental advantages, progress, and challenges associated with this innovative approach. Furthermore, it delves into the integration of functionalized nanostructured semi-metal-based compounds as lubricant additives in non-edible vegetable oils, paving the way for improved tribological properties. We anticipate that this book will serve as a catalyst for stimulating the interests of scholars and researchers within the academic sphere, igniting a new wave of exploration and inquiry into the realm of nanolubricants. Through the dissemination of the latest advancements in lubrication science, this volume aims to address pressing concerns surrounding economic feasibility, environmental acceptability, sustainability, and overall viability.

We extend our gratitude to the contributors who have shared their expertise, and we invite all readers to join us on this enlightening expedition into the world of nanolubricants. We acknowledge the great efforts of the eminent authors without whom this book would have been unimaginable. We also appreciate the interest shown and the support given by the dedicated team at Scrivener Publishing, especially Martin Scrivener, which allowed us to compile this handbook.

Dr. Mohd Yusuf
Dr. Lalit Prasad
Dr. Shafat Ahmad Khan
February 2024

1

An Insight into Nanolubrication and Nanolubricants

Deepshikha Singh[1,2], Wasim Khan[1†] and Mohd Yusuf[3*]

[1]*Department of Petroleum Engineering, School of Science and Technology, Glocal University, Mirzapur Pole, Saharanpur, Uttar Pradesh, India*
[2]*Department of Chemical Engineering, Madan Mohan Malaviya University of Technology, Gorakhpur, Uttar Pradesh, India*
[3]*Department of Natural and Applied Sciences, School of Science and Technology, Glocal University, Mirzapur Pole, Saharanpur, Uttar Pradesh, India*

Abstract
Over the last decade, nanomaterials and nanotechnology have been increasingly used in lubrication to prepare anti-friction mediums in industry-relevant tribological applications. Various nanolubricants are being developed as viable replacements for standard lubricants containing chlorine, phosphorus and sulfur, which pollute the environment when discarded. This chapter provides an overview of recent advances in nanolubricants, focusing on using nanoadditives in lubricant preparation. Finally, the applications of nanolubricants are summarized, along with future research directions.

Keywords: Nanomaterials, anti-friction medium, nanoadditives, tribological applications

1.1 Introduction

The use of nanotechnology in lubrication refers to nanolubrication, one of the most recent innovative technologies used in lubrication science, to provide an appropriate solution for energy loss. When moving materials come into contact, friction and wear occur, and studying these phenomena

**Corresponding author*: yusuf1020@gmail.com
†*Corresponding author*: wsmkhan03@gmail.com

Mohd Yusuf, Lalit Prasad and Shafat Ahmad Khan (eds.) Nanolubricants: Generation and Applications, (1–16) © 2024 Scrivener Publishing LLC

is critical in many applied sciences [1, 2]. It is now critical to develop lubricants that can be used in harsh environments, prompting research into the use of nanoparticles as lubricant additives. In terms of development, nanoparticles are the most recent lubricant additives. Base oil (lubricant), nanoadditives, and surfactants are the three main components of nano-based lubricants. Nanoparticles are colloidal solid particles in lubricating oil [3–6]. Efforts have been proposed to incorporate various nanoadditives into base lubricants to develop suitable nanolubricants to improve lubricant friction, wear resistance, and extreme pressure properties [7–11]. Nanolubricants have a wide range of applications, including industrial, automotive, aerospace, and biomedical. In the automotive industry, nanolubricants can be used to improve fuel efficiency, reduce emissions, and extend the life of engine components. In the aerospace industry, nanolubricants are used to improve the performance of aircraft engines and other critical components. Nanolubricants are a growing field, with continued research and development to improve their properties and expand their applications. The development of nanolubricants is driven by the need for more sustainable and energy-efficient lubrication solutions and the increasing demand for high-performance lubricants in various industries.

Nanoscale particles in lubricants have shown great promise in improving tribological properties and as a result, energy conservation. This chapter aims to provide the information needed to comprehend the fundamentals of nanolubrication, the mechanisms involved, and the forces at the nanoscale. Nonetheless, rising demand and resource depletion force us to take the steps necessary to preserve and conserve energy. This chapter investigates recent advances and developments in nanolubricants that may serve as a responsive outlook for future research directions.

1.2 Advantages of Nanolubricants

Nanolubricants are used widely because they have many advantages over traditional lubricants. These advantages include the following:

- Nanoparticle-laden lubricants have higher surface-to-volume ratios and better stability than lubricants dispersed with microsized particles because the added nanoparticles become more chemically reactive due to the presence of a large number of surface atoms, which aids in surface modification of the nanoparticles for stable dispersion in conventional lubricants.

- The risks and toxicity caused by adding additives to increase lubricant qualities are largely minimized by employing environmentally safe nanoparticles (NP).
- Nanoparticles included in lubricants may enter the gap between the tribo-pair in action, causing wear to be reduced.
- Nanolubricants have improved thermophysical characteristics (particularly thermal conductivity), which help in the extraction of heat created by excessive friction and wear.

1.3 Preparation of Nanolubricants

Nanolubricants are prepared by one- and two-step methods, as shown in Figure 1.1. The one-step method for the preparation of nanolubricants involves the direct addition of nanoparticles to the base oil. This simple method can achieve a high concentration of nanoparticles in the lubricant. The one-step method is often favored for its ease of preparation, quick results, and reduced cost. The nanoparticles used in this method can be directly dispersed in the base oil through ultrasonication or mechanical stirring. However, this method can result in poor dispersion and aggregation of the nanoparticles, leading to reduced lubrication performance. Techniques in the one-step method include microwave irradiation [12], chemical reduction [13], laser ablation [14], polyol process [15],

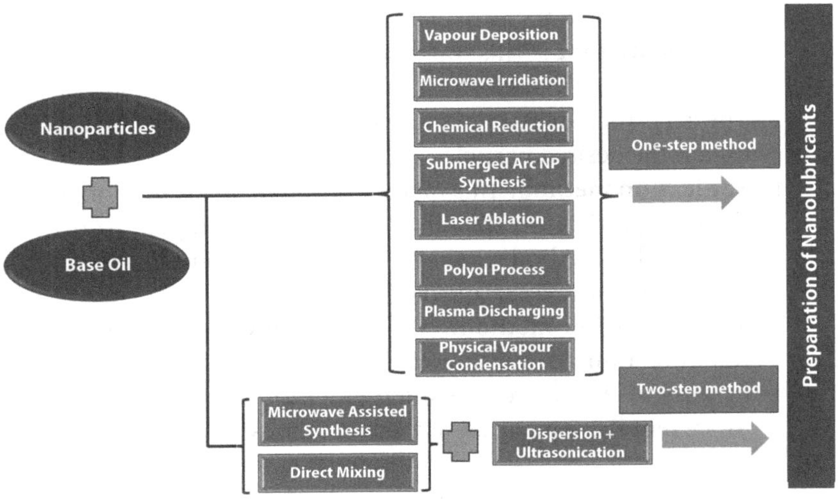

Figure 1.1 One-step and two-step preparation methods for nanolubricants.

physical vapor condensation method [16], plasma discharging technique [17], and submerged arc nanoparticles (NP) synthesis system [18].

The two-step method for preparing nanolubricants involves synthesizing the nanoparticles and adding them to the base oil. This method results in a more uniform nanoparticle distribution and higher purity than the one-step method. In the first step, the nanoparticles are synthesized through methods such as precipitation, co-precipitation, sol–gel, and hydrothermal synthesis. In the second step, the synthesized nanoparticles are added to the base oil and dispersed using ultrasonication or mechanical stirring techniques. This method results in better distribution and stability of the nanoparticles in the lubricant, leading to improved tribological properties.

1.3.1 Methods of Nanolubricant Preparation

There are several methods for the preparation of nanolubricants [19–21]; some of the common ones are

1. **Chemical reduction method:** This involves reducing metal ions to form metal nanoparticles, which can then be used as lubricant additives. The most commonly used metal nanoparticles are silver, copper and gold.
2. **Mechanical milling method:** This method involves grinding the lubricant base oil and the nanoparticle additive together using a ball mill, attritor mill, or other similar equipment. This method allows for the homogeneous dispersion of the nanoparticles in the base oil.
3. **Sol–gel method:** This method involves the synthesis of nanoparticles through the gelation of a precursor solution. The gel is then dried, and the resulting material is ground to obtain the desired particle size.
4. **Sonication method:** This method involves the dispersion of the nanoparticles in the lubricant base oil using ultrasonic energy. This method results in the homogeneous dispersion of the nanoparticles in the base oil.
5. **Microemulsion method:** This method involves the preparation of a microemulsion of the lubricant base oil and the nanoparticle additive. The microemulsion is then stabilized to prevent the separation of the two phases.

The preparation of nanolubricants can also be performed using various other methods, including the surfactant-free method, the surfactant addition method, and the surfactant-modified method. The surfactant-free method involves the direct addition of nanoparticles to the base oil without the use of surfactants (Figure 1.2). This method is simple, but it may result in poor dispersion and aggregation of the nanoparticles, leading to reduced lubrication performance. The surfactant addition method involves the addition of surfactants to the base oil to help disperse the nanoparticles. The surfactants reduce the surface tension between the base oil and the nanoparticles, improving the distribution and stability of the nanoparticles in the lubricant. This method can result in improved tribological properties, but using surfactants can lead to additional costs and environmental concerns. The surfactant-modified method involves the modification of the surface of the nanoparticles with surfactants to improve their dispersion in the base oil. This method results in improved stability and distribution of the nanoparticles in the lubricant, leading to improved tribological properties. However, this method requires additional steps in synthesizing the nanoparticles and can be more complex and costly than other methods.

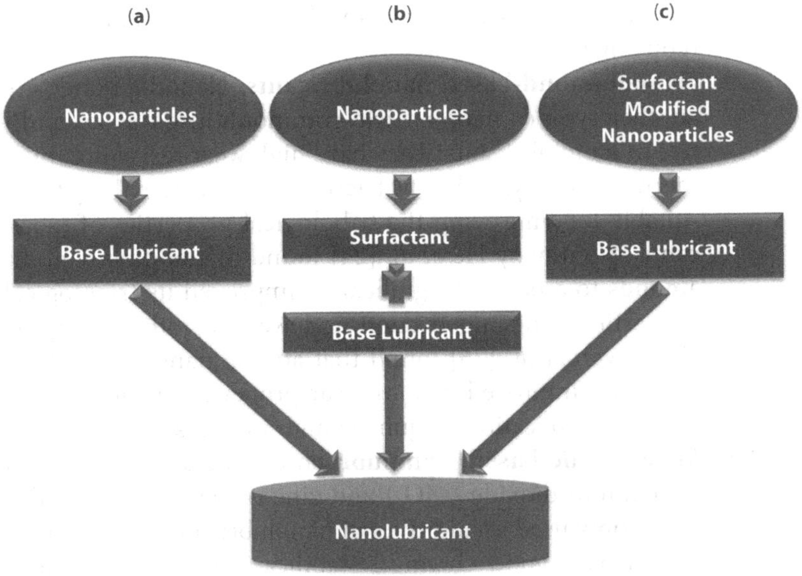

Figure 1.2 (a) Surfactant-free method, (b) surfactant addition method, and (c) surfactant-modified method.

1.3.2 Types of Nanolubricants Based on Additives' Characteristics

A literature review on the types of nanolubricants based on their additives' characteristics can provide an understanding of the various nanolubricant properties. Nanolubricants contain nanoparticles, which can improve their tribological properties such as reducing friction and wear. There are several types of nanolubricants based on the type of nanoparticle used as an additive. Carbon nanotubes, nanodiamonds, metal oxides, and graphene are some of the most commonly used nanoparticles in nanolubricants as shown in Table 1.1.

1. **Carbon nanotube (CNT)-based nanolubricants:** Carbon nanotubes are commonly used as additives in nanolubricants due to their high mechanical strength, high thermal conductivity and excellent wear resistance. These properties make CNTs suitable for reducing friction and wear in various industrial applications. A study by Ma *et al.* [22] found that adding multi-walled carbon nanotubes (MWCNTs) to a base oil significantly reduced friction and wear, compared to the base oil alone. Another study by Bhowmick *et al.* [23] found that adding single-walled carbon nanotubes (SWCNTs) to a base oil improved its tribological properties, reducing friction and wear.
2. **Nanodiamond-based nanolubricants:** Nanodiamonds are another type of nanoparticle commonly used as an additive in nanolubricants. They have high wear resistance, high thermal stability, and a low friction coefficient, making them suitable for improving the tribological properties of lubricants. A study by He *et al.* [24] found that adding nanodiamonds to a base oil significantly improved its tribological properties, reducing friction and wear. Another study by Vasudevan *et al.* [25] found that adding nanodiamonds to a base oil improved its anti-wear properties, which can be useful in applications requiring high wear resistance.
3. **Metal oxide-based nanolubricants:** Metal oxides, such as titanium dioxide (TiO_2) and zinc oxide (ZnO), are also commonly used as additives in nanolubricants. These metal oxides have good chemical stability, high thermal stability and low friction coefficients, making them suitable for improving the tribological properties of lubricants. A study

Table 1.1 Types of nanolubricant based on additives, with examples and uses.

Type of nanolubricant	Additives	Particle size	Examples	Uses
Carbon-based	Carbon nanotubes (CNTs), graphene	1–10 nm	Multi-walled carbon nanotubes (MWCNTs), single-walled carbon nanotubes (SWCNTs), graphene/oil	Low friction, high wear resistance, high thermal stability
Metal oxide-based	Titanium dioxide (TiO_2), zinc oxide (ZnO)	5–50 nm	TiO_2/oil, ZnO/oil, aluminum oxide (Al_2O_3/oil)	Improved anti-wear properties, high thermal stability, improved oxidation resistance
Diamond-based	Nanodiamonds	2–10 nm	Nanodiamond/oil, cubic boron nitride (cBN)/oil	High wear resistance, high thermal stability, low friction
Hybrid	Combination of different types of additives	Varies	Graphene/TiO_2/oil, CNTs/ZnO/oil	Improved wear resistance, reduced friction, high thermal stability
Silica-based	Silica (SiO_2)	10–50 nm	SiO_2/oil	Improved anti-wear properties, improved oxidation resistance, high thermal stability
Metal-based	Metal nanoparticles	10–50 nm	Au/oil, Ag/oil	Improved anti-wear properties, high thermal stability, improved oxidation resistance

by Li *et al.* [26] found that adding TiO_2 nanoparticles to a base oil significantly improved its tribological properties, reducing friction and wear. Another study by Jang *et al.* [27] found that adding ZnO nanoparticles to a base oil improved its anti-wear properties, which can be useful in applications requiring high wear resistance.

4. **Graphene-based nanolubricants:** Graphene is another material used as an additive in nanolubricants due to its high mechanical strength, high thermal conductivity and low friction coefficient. A study by Li *et al.* [28] found that adding graphene to a base oil significantly improved its tribological properties, reducing friction and wear. Another study by Song *et al.* [29] found that adding graphene to a base oil improved its anti-wear properties, which can be helpful in applications requiring high wear resistance.

1.4 Lubrication Mechanism

Lubrication is a crucial process in mechanical systems that helps to reduce friction and wear between moving surfaces, thus extending the lifespan and improving the efficiency of the machinery. Nanoparticles are added to the lubricant to improve its tribological properties via two lubricating mechanisms: direct and indirect lubrication mechanisms. Ball-bearing and protective/tribofilm mechanisms are direct lubrication mechanisms, whereas mending and polishing mechanisms are indirect lubrication mechanisms. Several lubrication mechanisms reduce friction and wear, including boundary, mixed, and full-film lubrication [30–37]. The following are the different mechanisms of nanoparticle lubrication:

1. **Boundary lubrication:** This type of lubrication occurs when the surfaces in contact are separated by a thin film of lubricant that is insufficient to separate the surfaces. The friction and wear in this regime are dominated by direct metal-to-metal contact between the surfaces. To improve boundary lubrication, additives such as extreme pressure agents are added to the lubricant to enhance its wear protection properties.
2. **Mixed lubrication:** This is a combination of boundary and full-film lubrication, where a thin layer of lubricant separates

the surfaces, but the lubricant film is not thick enough to completely separate the surfaces. In this regime, both metal-to-metal contact and hydrodynamic forces play a role in determining the friction and wear between the surfaces.

3. **Full-film lubrication:** This type of lubrication occurs when a thick lubricant film completely separates the surfaces in contact, preventing any direct metal-to-metal contact. Hydrodynamic forces, such as pressure gradients, shear forces, and lubricant viscosity, determine this regime's friction and wear.
4. **Physical interlocking:** In this mechanism, the nanoparticles interlock physically with the asperities on the surfaces, reducing the contact area between the surfaces and thus reducing friction.
5. **Chemical interlocking:** This mechanism involves the formation of chemical bonds between the nanoparticles and the surfaces, which locks the nanoparticles in place and reduces the possibility of metal-to-metal contact.
6. **Load transfer:** The small size of the nanoparticles enables them to transfer the load from one surface to another, reducing the pressure on any one surface, and thereby reducing friction and wear.
7. **Surface modification:** Nanoparticles can also modify the surface properties of the surfaces in contact, reducing the friction and wear between them.
8. **Film formation:** The presence of nanoparticles can alter the rheological properties of the lubricant, leading to the formation of a more robust and stable lubricant film that separates the surfaces and reduces friction and wear.

1.5 Tribological and Thermophysical Properties of Nanolubricants

Tribology is the science and engineering of interacting surfaces in relative motion, including friction, wear, and lubrication. Nanolubricants, which are lubricants containing nanoparticles, have been developed in recent years to improve the tribological properties of conventional lubricants. The following are the tribological and thermophysical properties of nanolubricants [38, 39]:

1.5.1 Tribological Properties

1. **Friction reduction:** Nanolubricants have been proven to dramatically reduce friction compared to traditional lubricants due to the tiny size of the nanoparticles, which allows them to penetrate the interface between moving surfaces and minimize the contact area. Friction reduction can contribute to higher efficiency, longer mechanical component lifespan, and lower energy usage. Non-polar functional group-containing hybrid and mono-nanoparticle-based nanolubricants showed improved friction and wear resistance. Due to the increased antioxidant nature of the nanolubricant, oleic acid and oleylamine-based-modified nanostructures have a synergistic impact on friction and wear reduction. Friction and wear reduction in non-polar (metal, carbon)-based nanolubricants are greater than metal oxide-based nanolubricants due to the increasing polar (Al-O, Ni-O, Ti-O, O-H) function group in nanolubricants. Nanoparticles having a larger surface-to-volume ratio and a lower concentration have a superior tribological impact. Carbon nanostructures, in particular, have exceptional tribological characteristics due to their conducting, semiconducting, and insulating nature [39].
2. **Wear reduction:** The presence of nanoparticles in the lubricant can also reduce wear by reducing the friction between the surfaces and preventing metal-to-metal contact. The small size of the nanoparticles enables them to physically interlock with the asperities on the surfaces, reducing the contact area and thus reducing wear.
3. **Load-carrying capacity:** Nanolubricants have been shown to have improved load-carrying capacity compared to conventional lubricants due to the ability of the nanoparticles to transfer the load from one surface to another. This reduces the pressure on any one surface and improves the overall performance of the lubricant.
4. **Thermal stability:** The thermal stability of nanolubricants can be improved compared to conventional lubricants due to the presence of nanoparticles. The nanoparticles can act as heat sinks, absorbing heat from the system and reducing the temperature of the lubricant, reducing the degradation

of the lubricant and improving its overall performance. A nanolubricant with higher surface-to-volume ratio and optimum concentrations of nanoparticle exhibits excellent stability. Oleic acid and oleylamine-based non-polar surfactant show better stability in lubricant medium due to reduced agglomeration (intermolecular forces) between the nanoparticles. Specifically, carbon-based nanostructures showed better stability than other nanostructures due to similar functional groups.

1.5.2 Thermophysical Properties

1. Viscosity

The viscosity of nanolubricants can be modified compared to conventional lubricants, depending on the type and size of the nanoparticles used. Viscosity can be increased or decreased based on working temperature and nanostructures concentration, which can affect the ability of the lubricant to flow and the thickness of the lubricant film that separates the surfaces. Stable aggregation or mono-nanoparticles help to enhance lubricant viscosity. Specifically, the non-polar functional group with better surface area contained nanoparticles that helps to improve the viscosity index in the base lubricant.

2. Surface tension

The surface tension of nanolubricants can be modified compared to conventional lubricants due to the presence of the nanoparticles. Surface tension can be increased or decreased, which can affect the ability of the lubricant to wet the surfaces and form a stable lubricant film. One important consideration in understanding the surface tension of nanolubricants is the interfacial behavior of the particles themselves. Due to their small size, nanolubricant particles have a large surface area-to-volume ratio, which can make them highly reactive and prone to aggregation. To stabilize the particles and prevent them from clumping together, surface modifiers and dispersants are often used. Another factor that can affect the surface tension of nanolubricants is the presence of surface-active agents, which are compounds that can adsorb to the surface of the particles and alter their interfacial behavior. These agents can reduce the surface tension of the nanolubricant, making it easier for the lubricant to spread and form a thin film on the surface on which it is applied [40].

3. Density

The density of a nanolubricant is an important physical property that affects its flow behavior, lubrication efficiency, and storage requirements. The density can be increased or decreased compared to conventional lubricants, depending on the type and size of the nanoparticles used, which can affect the ability of the lubricant to flow and its overall performance. For example, metal nanoparticles are generally denser than most base fluids, and their addition to the fluid can increase the overall density of the nanolubricant. Conversely, carbon nanotubes have a lower density than most base fluids, and their addition can decrease the overall density of the nanolubricant.

4. Thermal conductivity

The thermal conductivity of nanolubricants increases with temperature. High interparticle spacing between nanoparticles is a symptom of poor thermal conductivity. Metal, metal oxide, carbon (conducting nanostructures), and hybrid nanoparticles have higher thermal conductivities and a higher surface-to-volume ratio, making them ideal for use as a nanolubricant. Discrete nanostructures with high surface area or stable aggregation boost thermal conduction processes even further. Nanolubricants offer better thermal conductivity due to more frequent nanoparticle aggregation and improved suspension characteristics although there is no association between enhancing thermal conductivity and other lubricant properties such as friction/wear, kinematic viscosity, flash point, and pour point.

5. Flash point

The flash point of nanolubricants refers to the temperature at which the lubricant will ignite when exposed to a flame or spark. Nanolubricants can have different flash points than traditional lubricants due to their unique chemical properties and composition. The addition of nanoparticles can increase the flash point of lubricants, making them safer to use in high-temperature applications. The improved thermal stability of nanolubricants is attributed to the ability of nanoparticles to act as heat sinks and reduce the chances of the lubricant reaching its ignition temperature. Flashpoint is higher for non-polar than polar nanoparticles and increases with the concentration of nanoparticles.

6. Pour point

The pour point of a nanolubricant is a critical property that influences its ability to flow and lubricate under low-temperature conditions. The pour point of a nanolubricant is determined by the base fluid and the type and concentration of nanomaterials used. Nanomaterials can modify the base fluid's pour point by adsorbing onto the fluid's surface and altering its physical properties. For example, metal nanoparticles can increase the viscosity of the base fluid, which can lead to an increase in the pour point. However, carbon nanotubes can decrease the pour point by forming a stable network within the fluid that prevents the formation of ice crystals. Non-polar (C–H, C=C=C, C=C, C≡C) functional group containing carbon nanostructures shows a higher pour point [41].

1.6 Conclusion and Future Directions

The actual behavior of a lubrication system can be complex and can vary based on many factors, including the geometry of the surfaces, the speed and load conditions, the temperature and pressure of the system, and the properties of the lubricant itself. The actual performance of nanoparticle lubrication can be influenced by several factors, including the size and shape of the nanoparticles, the properties of the surfaces, the temperature and pressure of the system, and the properties of the lubricant. In conclusion, nanolubricants have been shown to have improved tribological and thermophysical properties compared to conventional lubricants. The small size of the nanoparticles enables them to penetrate the interface between moving surfaces and reduce friction, wear, and temperature. The tribological and thermophysical properties of nanolubricants can be tailored to suit the specific needs of a particular application by selecting the type and size of the nanoparticles used.

References

1. Kechar, M., Boulon, G., Bouzid, J., Nano-lubricants for energy efficiency: A review. *Energy*, 156, 943–961, 2018a.
2. Kechar, M., Boulon, G., Bouzid, J., Nanoparticle-based lubricants: A review of the current status and perspectives. *Tribol. Int.*, 121, 121–139, 2018b.
3. Zhang, H.J., Song, Y.X., Li, Y.Y., Guo, Y.X., Nano lubricants: Types, synthesis and applications. *J. Ind. Eng. Chem.*, 43, 165–173, 2017.
4. Gao, L., Liu, B., Sun, D., A review on tribological properties of nanoparticle-added lubricants. *Wear*, 376–377, 99-114, 2017.
5. Shinoda, K. and Ikura, Y., Nanoparticle lubrication: A review. *J. Tribol.*, 129, 1, 1–8, 2007.

6. Ge, Y. and Lin, J., Preparation and tribological properties of nanolubricants. *Tribol. Lett.*, 66, 1, 56, 2018.
7. Wang, Q., Jia, X., Li, Y., Preparation and tribological properties of TiO_2 nanoparticle-doped engine oil. *Ind. Eng. Chem. Res.*, 55, 11, 2979–2985, 2016.
8. Li, J., Li, Y., Li, C., Li, H., Du, Y., Preparation and performance of Al_2O_3 nanoparticle-doped engine oil. *J. Ind. Eng. Chem.*, 49, 97–102, 2017.
9. Wang, Q., Jia, X., Li, Y., Preparation and tribological performance of ZnO nanoparticle-doped engine oil. *Wear*, 340–341, 116-121, 2015.
10. Kim, Y.S., Kim, Y.J., Kim, J.K., Preparation and tribological properties of Ag nanoparticle-doped engine oil. *Tribol. Int.*, 89, 201–207, 2015.
11. Zhang, X., Fang, X., Liu, Y., Dai, L., Preparation and tribological properties of MoS2 nanoparticle-doped engine oil. *Tribol. Trans.*, 59, 6, 581–588, 2016.
12. Wang, H., Xu, J.-Z., Zhu, J.-J., Chen, H.-Y., Preparation of CuO nanoparticles by microwave irradiation. *J. Cryst. Growth*, 244, 88–94, 2002.
13. Sandhya, S.U. and Nityananda, S.A., A facile one step solution route to synthesize cuprous oxide nanofluid. *Nanomater. Nanotechnol.*, 3, 5, 2013.
14. Han, S.Y., Shin, S.Y., Lee, H.-J., Lee, B.-J., Lee, S., Kim, N.J., Kwak, J.-H., Effects of annealing temperature on microstructure and tensile properties in ferritic lightweight steels. *Metall. Mater. Trans. A Phys. Metall. Mater. Sci.*, 43, 843–853, 2012.
15. Sun, Y., Mayers, B., Herricks, T., Xia, Y., Polyol synthesis of uniform silver nanowires: A plausible growth mechanism and the supporting evidence. *Nano Lett.*, 3, 955–960, 2003.
16. Eastman, J.A., Choi, S.U.S., Li, S., Yu, W., Thompson, L.J., Anomalously increased effective thermal conductivities of ethylene glycol-based nanofluids containing copper nanoparticles. *Appl. Phys. Lett.*, 78, 718–720, 2001.
17. Hsin, Y.L., Hwang, K.C., Chen, F.-R., Kai, J.-J., Production and *in-situ* metal filling of carbon nanotubes in water. *Adv. Mater.*, 13, 830–833, 2001.
18. Lo, C.H., Tsung, T.T., Chen, L.C., Shape-controlled synthesis of Cu-based nanofluid using submerged arc nanoparticle synthesis system (SANSS). *J. Cryst. Growth*, 277, 636–642, 2005.
19. Hlaing, T. and Phung, T.V., A review on the synthesis methods of nanolubricants. *J. Cleaner Prod.*, 169, 143–155, 2018.
20. Shukla, A.K. and Jha, R.K., Nanoparticle-based lubricants: A review of the synthesis methods and applications. *J. Ind. Eng. Chem.*, 47, 168–184, 2017.
21. Morshed, A., Wu, H., Jiang, Z., A comprehensive review of waterbased nanolubricants. *Lubricants*, 9, 89, 2021.
22. Ma, Q., Li, J., Sun, Y., Multi-walled carbon nanotubes as friction and wear reducing additive in lubricants. *Wear*, 271, 1-2, 94–99, 2012.
23. Bhowmick, A.K., Wadhwa, R., Roy, S., Friction and wear behaviour of single-walled carbon nanotube (SWCNT) reinforced nanolubricants. *Wear*, 334-335, 27–35, 2015.

24. He, X., Zhang, Y., Liu, X., Study on the friction and wear behaviour of nanodiamond/base oil composites. *Wear*, 269, 1-2, 267–274, 2010.
25. Vasudevan, S. and Ajayan, P.M., Nanodiamond reinforced lubricant for tribological applications. *Tribol. Int.*, 86, 129–134, 2015.
26. Li, Z., Yang, Z., Song, Y., Preparation and tribological properties of titanium dioxide nanoparticle/oil composite lubricants. *Appl. Surf. Sci.*, 280, 92–96, 2013.
27. Jang, J.H., Kim, J.H., Kim, J.K., Kim, K.S., Synthesis and tribological properties of zinc oxide nanoparticle/oil composite lubricants. *J. Ind. Eng. Chem.*, 21, 1282–1286, 2015.
28. Li, Y., Wang, J., Xia, Y., Tribological behaviour of graphene-reinforced lubricating oil under high pressure and high temperature. *Wear*, 282–283, 44-48, 2012.
29. Song, X., Zhang, Z., Wang, H., Wang, X., Synthesis and tribological properties of graphene/oil composite lubricants. *Appl. Surf. Sci.*, 355, 1021–1026, 2015.
30. Lubrecht, A.A., *Lubrication Fundamentals, Tribology Series*, vol. 16, Elsevier, Amsterdam, 1985.
31. French, R.H. and Wallin, M.P., *Tribology-Friction and Wear of Engineering Materials*, Springer, Singapore, 2014.
32. Clayton, J.M. and MacKay, D.J.C., *Lubrication and Wear of Engine Components*, Society of Automotive Engineers, Inc, Warrendale, PA, USA, 1990.
33. Michal, G.H., *Lubricants and Lubrication*, Elsevier Science, Amsterdam, 2006.
34. Shinoda, K. and Ikura, Y., Nanoparticle lubrication: A review. *J. Tribol.*, 129, 1, 1–8, 2007.
35. Smith, C.A. and Tribus, J.A., Nanoparticle lubrication: A review of the literature. *Tribol. Trans.*, 53, 3, 441–450, 2010.
36. Zhang, H.J., Song, Y.X., Chen, W.Q., Nanoparticle lubrication: A review. *J. Nanopart. Res.*, 14, 1, 767–782, 2012.
37. Lu, Y. and Wang, Q., Nanoparticle-enabled lubrication: A review of fundamentals, progress, and applications. *Tribol. Int.*, 113, 140–152, 2017.
38. Smith, C.A. and Tribus, J.A., Nanoparticle lubrication: A review of the literature. *Tribol. Trans.*, 53, 3, 441–450, 2010.
39. Pownraj, C. and Valan Arasu, A., Effect of dispersing single and hybrid nanoparticles on tribological, thermo-physical, and stability characteristics of lubricants: A review. *J. Therm. Anal. Calorim.*, 143, 1773–1809, 2021.
40. Rahman, M.M. and Guo, Y., Surface Tension of Nanolubricants: A Review. *Tribol. Trans.*, 62, 6, 1028–1043, 2019.
41. Saeedi, A., Hamdi, H., Najafi, G., Nanolubricants: A review of the recent advances in research on their preparation and properties. *Tribol. Int.*, 110, 250–267, 2017.

2

Nanolubrication Chemistry and Its Application

Smrita Singh[1]*, Ashutosh Singh Chauhan[2] and Lalit Prasad[3]†

[1]*Creative Bioinformatics and Science, Morna, Bijnor, Uttar Pradesh, India*
[2]*Archaeological Survey of India, Agra Fort, Agra, Uttar Pradesh, India*
[3]*Galgotias University, Greater Noida, Uttar Pradesh, India*

Abstract

Lubrication is a method to reduce friction in moving machinelike parts. However, rising demand and ongoing resource depletion urge us to take essential steps to save and conserve energy. Nanolubrication is a viable option for reducing energy consumption. Nanolubricant films must be resistant to adhesives, cohesiveness, and stresses in order to be effective. The way these films is organized, whether they are single molecules or mixed species, has a big impact on how well they perform. A film's most essential property is its ability to adhere to a surface. Physical or chemical adsorption is how molecules adhere to a surface. Major factors in affecting molecular weight molecules are bond strength, nature of the bonds, the number of bonds per molecule, molecular orientation, and packing density per unit area. Molecular adhesion to a surface can be homogenous and equally distributed on an ideal surface. The crystalline orientations and phases present at the surface of silicon have an impact on its chemical reactivity and mechanical capabilities. Dangling bonds are formed when the surface is mechanically damaged or thermally stimulated, producing high-energy active sites for adsorption and contact. Surface layers of silicon nitride and silicon carbide can be produced on silicon using gas phases in a vacuum. Nickel and diamond-like carbon are the other two materials commonly used in microsystems. With magnetic lubrication of hard drives, the hard drive is protected by a monolayer of perfluoropolyether (PFPE). Commonly used PFPE has two OH functional groups at the ends of molecules. Colloid physics and chemistry offer exciting opportunities for research to reduce friction and wear in various metallurgical contexts. Different chemical methods

**Corresponding author*: smrita1@gmail.com
†*Corresponding author*: josh.iitd81@gmail.com

Mohd Yusuf, Lalit Prasad and Shafat Ahmad Khan (eds.) Nanolubricants: Generation and Applications, (17–60) © 2024 Scrivener Publishing LLC

have been developed to produce nanoparticles, permitting the development of an oil-stable dispersion of selected efficient tribochemical film precursors.

Keywords: Perfluoropolyether, tribochemical film, packing density, molecular orientation, colloid, nanoparticles

2.1 Introduction

One of the most effective ways to reduce energy use and increase the productivity of machinery used in various industries has been thought to be lubrication. By introducing a lubricant or lubricating agent between two surfaces that are moving in close proximity to one another, lubrication is described as the process or technique used to reduce friction and wear [1]. A slippery film is created when a lubricant is present, significantly reducing friction, wear, and tears between mating surfaces. In addition to reducing friction and wear, lubricants also enhance cooling and heat transfer, reduce vibration and noise, remove dirt and pollutants on their own, and, in some situations, prevent leaks. Lubricants should have properties like high boiling point, acceptable viscosity, low freezing point, high thermal stability, and resistance to oxidation and corrosion in order to fulfill these functions. The tribological characteristics of the lubricating media are crucial for effective machine lubrication. Mineral oils have been historically used to lubricate tribo-pairs effectively on the global market. The various chemical modifications, sometimes referred to as additives, improve the tribological effectiveness of the lubricants. The additives function as viscosity index enhancers [2], anti-wear agents, extreme pressure agents, anti-corrosion agents, and friction modifiers. High temperatures and pressures have been observed to cause the standard additives to fail. These additives also have a number of technical flaws, including inherent toxicity, non-biodegradability, high reactivity, ineffectiveness at low temperatures, and slow speeds, among others [3]. This has created a need for environmentally friendly lubricants that will improve tribo-pair performance while being less harmful to the environment. With the development of nanotechnology, a wide range of different types of tailored materials are now readily available, making the answer simple. Different kinds of nanoparticles are perfect lubricant additions because they do not have the drawbacks of traditional lubricants listed above. Therefore, adding extremely modest amounts of nanoparticles to lubricants may improve their tribological qualities. Generally speaking, such tailored fluids with nanoparticle dispersions are referred to as nanofluids [4–7]. However, "nanolubricants" have been used frequently in the literature [2, 8–12] to refer to nanofluids used as lubricating media. More specifically, nanolubricants are stable colloidal suspensions of nanosized materials (such as nanofibers, nanotubes, nanowires, nanorods, and nanosheets) that are added in extremely small amounts (<1 wt%) to traditional lubricants to improve their tribological properties.

2.2 Nanolubrication and Its Requirements

Controlling and maintaining the surface energy state—defined by electrostatic charges, defects, and active sites—is necessary to safeguard the surface. Due to the lack of regulated contact geometry and insufficient molecules to produce such films, the load is unlikely to be maintained at the nanoscale by the macroscale squeezed-film mechanism. Since it is challenging to continually provide the molecules to replace the film, and because it is challenging to dispose of the degradation products, the nano-sacrificial film method is likewise not very likely. Therefore, a new lubricating principle is required. Tenacious, long-lasting surface coatings will be required to lubricate surfaces at the nanoscale. With the limited supply of lubricants, the film's effectiveness in reducing shear stress is very important as only one or two layers of molecules are available for the duration of the device's lifetime. The ability of the film to withstand repeated contact over a long period of time is also critical. Earlier, it was proved that a monolayer of fatty acids on glass surfaces reduces friction as proved by experiments of Bowden and Tabor [2]. Due to repeated sliding, these monolayer films did not last long and the film failed as some of the molecules were removed from the surface by mechanical rubbing. The effectiveness of solid lubricant films such as MoS_2, graphite, and Teflon, should be checked before using them, as any severe abrasive wear can be caused by inorganic degradation products such as MoOx or carbon particles. Micro/nanoscale devices need to be operated in a clean environment. If there are outlets and scraping devices for the degradation products, such as roller bearings, these foils could be effective. Any organic film that can function effectively is able to repair and refill, and remains intact when impacted or damaged. An important characteristic of a lubricant is self-repair mechanism due to the migration of molecules when lubricant molecules are removed from one site, due to contact, oxidation, or evaporation. Temperatures can rise at typically high speed contacts; therefore, in order to overcome this effect, these molecules must withstand thermal decomposition and oxidation. Low vapour pressure, volatility and high resistance to oxidation and decomposition is required for these lubricants. Perfluoropolyether (PFPE) is used as a lubricant in magnetic head-disk with a thickness of 1 and 2 nm. The main characteristics of PFPE are resistance to oxidation, low volatility (106 torr), and adherence to the surface by hydrogen bonding to withstand the high centrifugal force of disk rotation.

In summary, nanoscale lubrication (control of adhesion, stiction, and friction) requires lubricant molecules that should have the following important properties:

(1) non-volatile,
(2) resistant to oxidation,
(3) thermal degradation,
(4) good adhesion and cohesion with the surface, and
(5) self-repairing or regenerating.

A nanolubricant consists of three components (Figure 2.1) [13]:

(1) base oils: mineral oil, synthetic oil, and vegetable oil;
(2) additives: chemical additives and nano-additives; and
(3) surfactants.

Mineral oils, synthetic oils, and vegetable oils are the three types of base oils. Chemical additives and nano-additives are two more categories of additives that improve the fundamental tribological characteristics of oil [14]. Surfactants stabilize the nanolubrication system.

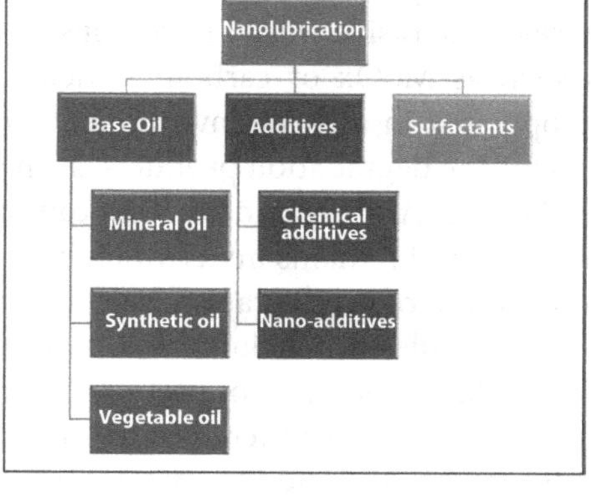

Figure 2.1 An overview of nanolubrication components.

2.3 Synthesis of Nanoparticles

There has been a variety of evolved nanoparticles, including metals, metal oxides, and sulfides (tungsten disulfide, molybdenum disulfide, and copper monosulfide), carbon-based nanoparticles (diamond, graphene, and fullerene), and nanocomposites (Cu/SiO_2, Cu/graphene oxide, Al_2O_3/SiO_2, Al_2O_3/TiO_2, and serpentine/La$(OH)_3$). Examples of metal nanoparticles are iron, silver, copper, gold, and zinc. ZnO, TiO_2, SiO_2, CuO, Fe_3O_4, ZrO_2, and Al_2O_3 are metal oxide nanoparticles. There are three methods for nanoparticle synthesis [1]:

(1) physical methods,
(2) chemical methods, and
(3) biosynthetic methods.

2.3.1 Physical Method

In order to create nanoparticles, the physical process of nanoparticle creation employs the use of mechanical, compressive, thermal, and electrical radiation. Ball milling, inert gas condensation, electrospray, pulsed evaporation, melt mixing, laser pyrolysis, flash spray pyrolysis, and other physical techniques are among them (Table 2.1) [1, 15].

Table 2.1 Physical methods for synthesis of nanolubricants.

S. no.	Physical method	Material used	Nanoparticle	Size (nm)	Reference
1	Ball milling	Graphite SiO_2	Si	5-500	[19]
		Graphite/Nitrogen	Nitrogen doped Carbon	210	[20]
		Iron powder	Magnetite	12-20	[21]
2	High energy ball milling	Iron sulphur	FeS_2	10	[22]
3	Inert condensation	------	Au Pd	1-5	[23]
		Manganese granules	Mn	2-20	[24]

(Continued)

Table 2.1 Physical methods for synthesis of nanolubricants. (*Continued*)

S. no.	Physical method	Material used	Nanoparticle	Size (nm)	Reference
4	PVD	HOPG, gold	Au	2-15	[25]
5	PVD Laser ablation	Zinc	ZnO_2	5-19	[26]
6	PVD/Sputtering	------	Cu-Au	Less than 7	[27]
7	Laser pyrolysis	Iron compound	Fe-FeO_2	14	[28]
		Titanium chloridepyTolyss ($FeCo_5$)	TiO_2/Fe-TiO_2		[29]
8	Spray pyrolysis	Magnesium nitrate/ Hexa methylene tetramine	MgO	9	[30]

Source: Adopted from Shafi WK, Charoo M. Nanolubrication systems: an overview. *Mater Today: Proc.* 2018;5(9):20621–20630.

2.3.2 Chemical Methods

This is the most popular and straightforward method for creating nanoparticles. Through a chemical reaction involving several precursors and solvents, the creation of nanoparticles involves their breakdown into atoms, followed by their aggregation into clusters [16]. Precursor concentration, pH, calcination temperature, and other factors all affect the size, shape, and stability [17]. Simple precipitation method, sol–gel method, hydrothermal, solvothermal, chemical vapor deposition, and polyol synthesis among others, are a few of the different chemical processes used to create nanoparticles. Table 2.2 provides a collection of several research publications on chemical synthesis techniques [1].

2.3.3 Biological Methods

The safest, least poisonous, and most environmentally responsible approach to creating nanoparticles is through biological processes. For the synthesis of nanoparticles, biosynthetic processes use plant extracts, bacteria, viruses, and fungi [18]. Research on the biosynthetic processes are depicted in Figures 2.1 and 2.2 (Table 2.3).

Table 2.2 Chemical methods for synthesis of nanolubricants.

S. no.	Chemical method	Precursor/Solvent	Nanoparticle	Size (nm)	Reference
1	Hydrothermal	Zirconia /NaOH	ZrO_2	15-30	[31]
		Zinc nitrate/ NaOH, DI water	ZnO	50	[32]
2	Solvothermal	Nickel/acetyl acetone/2-Butanone	NiO	5.5-6.5	[33]
3	Simple aqueous route	Zinc chloride/ Distilled water	ZnO	8-30	[34]
		$TiOSO_4$, /H_2O_2:	TiO_2	28-50	[35]
4	Sol-gel	Zirconium-n-Propoxide /n-Propanol, Water/ Ammonia	ZrO_2	30-100	[36]
		Titanium- iso-Propoxide isopropyl alcohol, water, NaOH, Pluronic (F-127)	TiO_2	10-12	[37]
5	Polyol	Silver nitrate ethelene glycol,	Silver	30-60	[38]
		PVP, NaCI	Palladiunm	2	[39]

Source: Adopted from Shafi WK, Charoo M. Nanolubrication systems: an overview. *Mater Today: Proc.* 2018;5(9):20621–20630.

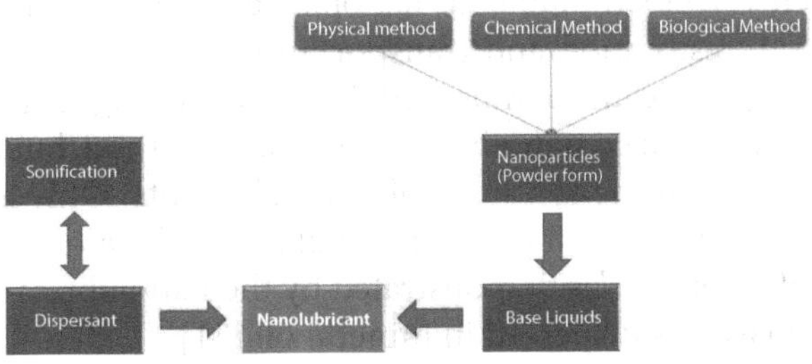

Figure 2.2 Synthesis of nanolubricant using two-step method.

Table 2.3 Biological methods for synthesis of nanolubricants.

	Biological method	Material used	Nanoparticle	Size (nm)	Reference
1	Plants extraction	Magnolia Kobus, Dropyros kaki $HIAuCl_4$	Gold	5-300	[40]
		Pine Ginkgo/ persimmon/ magnolia plantanaus/$AgNO_3$	Silver	5-500	[41]
2	Bacteria	Rhodopseudomonas capsutala, $HIAuCl_4$	Gold	10-20	[42]
		Gum karaya	CuO	4-8	[43]
3	Fungi	Aspergillus/$AgNO_3$	Silver	420	[44]
		Verticillium, $AuCl_4$	Gold	20-40	[45]
		Rhizoctoni, $AgNO_3$	Silver	20-50	[46]

Source: Adopted from Shafi WK, Charoo M. Nanolubrication systems: an overview. *Mater Today: Proc.* 2018;5(9):20621–20630.

2.4 Preparation of Nanofluids/Nanolubricants

The methods for preparing nanofluidics are the following [47]:

- one-step method and
- two-step method.

2.4.1 One-Step Method

This method involves simultaneous nanoparticle synthesis and dispersion. In this method, restoration and transportation are avoided, which minimizes clustering and leads to improved dispersion strength [1, 47]. Diverse research have synthesized nanofluids by one-step method, but not so much has been reported on the synthesis of nanolubricants. Only a few studies have demonstrated the synthesis of nanolubricants by a one-step process. One month of stability was demonstrated by the nanolubricant composed of silver nanoparticles dispersed in mineral oil [51].

2.4.2 Two-Step Method

Due to its affordability, this technique is the one most frequently employed to create nanofluids. Different physical, chemical, or biological techniques are used to manufacture powder-form nanoparticles of various sizes and morphologies first. The produced nanopowders are subsequently mixed with or without additions into the basic liquids [47]. Utilizing tools such as an ultrasonic agitator, a homogenizer, and a magnetic force agitator, nanopowders and liquids are mixed (Figure 2.2). Various studies have been conducted on nanofluids/nanolubricants. Batez *et al.* mixed PAO_6 with ZnO nanoparticles and a dispersant using an ultrasonic probe [48]. Modified CuO nanoparticles were mixed with vegetable oil using an ultrasonic shaker (Koshy *et al.*) [49]. Surface modification was performed to reduce the agglomeration of nanolubricants. Padgurska *et al.* mixed nanoparticles of copper, iron, and cobalt into mineral oil. From a simple precipitation method, MoS_2 was synthesized, and in rapeseed oil, three types of MoS_2 were mixed [50].

2.4.2.1 Disadvantages of the Two-Step Method

The stability of nanoparticles in base liquids is the main disadvantage. Two-stage mixing leads to nanoparticle aggregation and cluster formation, which degrades the performance of nanofluids.

2.4.3 Dispersion of Nanoparticles in Lubricating Oils

How effectively nanoparticles are dispersed in oil determines their effectiveness on tribological performance. Due to the least mass collection of particles, nanoparticles cannot form clusters, and as a result, they easily penetrate the surface and get deposited or roll on contact surfaces, thereby reducing friction and wear [1, 10]. Dispersion of nanoparticles can be done by several methods such as ultrasound, probe sonication, chemical stirring, and magnetic stirring. While dispersing nanoparticles, care should be taken in order to avoid the aggregation of the nanoparticles. Longer mixing times improve the stability of nanoparticles in lubricants. This method, however, limits the use of nanolubricants due to particle separation after a few days.

This problem is solved by the following methods:

(1) using surfactants [47] and
(2) surface functionalization of nanoparticles [52].

Lubricating oils contain dispersants, which are surfactants. Nanoparticles and oil tend to be dispersed together by dispersants. Among different types of surfactants, selecting the right type of surfactant is a very important factor according to the type of oil. For non-polar liquids such as lubricating oils, non-ionic surfactants with hydrophilic– lipophilic balance (HLB) values below nine have proven to be very efficient. Surface functionalization is another method that can be used to enhance the stability of nanolubricants. To provide stability and integrity, surface properties of the nanoparticles are modified. For surface modification of nanoparticles, several organic agents and coupling agents can be used [52].

2.4.4 Interaction Forces in a Nanofluids/Nanolubricant System

Interaction forces in nanofluidic systems arise from interactions between nanoparticles and the surface and between the nanoparticles themselves. Forces between nanoparticles affect the performance of nanoparticles as strong interactions can cause aggregation and compromise their utility [1]. The reason for stability is indicated by Derjaguin-Landau-Verwey-Overbeek (DLVO).

Different types of assumptions of interacting forces have been made by different researchers. The main forces include

(1) Van der Waals forces,
(2) electrostatic forces, and
(3) capillary forces.

The above said forces exist between nanoparticles and surfaces, and between two nanoparticles.

Van der Waals force and the electrostatic force are the two main forces involved in nanolubrication systems among the above forces [53].

2.4.4.1 Van der Waals Forces

These forces are the weak electrostatic forces that arise from the continuous vibration of electrons. This creates positively and negatively charged regions within the molecule that may result in the attraction or repulsion of two nanoparticles.

Based on interactions, Van der Waals forces have three types (Illustration 2.1):

(1) Keesom force: The Keesom force is generated by the permanent dipole moment of polar molecules [54].

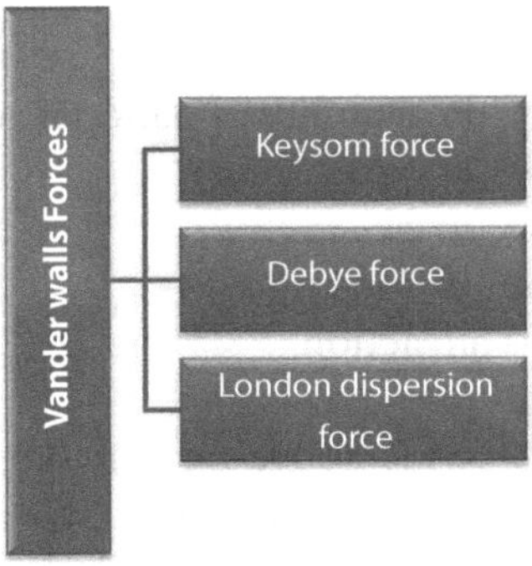

Illustration 2.1 Three types of Van der Waals forces.

(2) Debye force: Permanent dipoles may induce dipole moments in nearby molecules, and such interactions between molecules are known as Debye forces [55].
(3) London dispersion force: This is the weakest force and only interacts for a very short time. These forces are generated by transient dipole moments between molecules caused by the continuous motion of electrons [56].

Van der Waals interactions between identical materials are always fascinating as they can attract or repel between two dissimilar materials, depending on the medium in which they are dispersed.

2.4.4.2 Electrostatic or Electric Double Layer Force (EDL)

Nanoparticles suspended in a dielectric medium become charged on their surface. The surface charge of particles may be because of the dissociation or ionization of surface groups or the adsorption of ambient solution ions to the particle surface [57]. Detection of surface charge leads to polarization of the surrounding ion cloud. Positive charges start trapping negative ions and vice versa. This creates a charged layer around the particles. According to Stern Gouy Chapman's model, the outer layer is the diffusion layer, and the inner layer is the compact inner layer. The diffused outer layer is measured by the Debye screen length. The EDL thickness is known as the Debye length. The charge density of the system can be expressed as the Debye–Hückel charge density [1].

2.4.4.3 DLVO Theory

According to this theory, the stability of nanoparticles in liquids is dependent on the following two factors:

(1) electrostatic repulsion and
(2) Van der Waals attraction.

The combined effects of the above two factors are taken into account while predicting the stability of nanoparticles [47]. This theory defines that the interaction energy between two nanoparticles is equal to the sum total of the Van der Waals energy and the electrostatic energy.

DLVO theory **Van der Waals energy + electrostatic energy = interaction energy of two nanoparticles**

Particles are most efficiently dispersed when electrostatic repulsion forces dominate over Van der Waals interactions.

2.4.4.4 Capillary Forces

When a thin film of water forms on a hydrophilic surface and is adsorbed to the surface (which may be surfactant-modified nanoparticles or the surface itself) by the action of capillary forces, these forces acting as a lubricated system becomes very important [58]. Humidity may result in the formation of a thin film of water. As a result, the water film disassembles the surfactant-modified nanoparticles, minimizing the effect of the surfactant and causing dispersion problems. As a result, when the surfactant layer is replaced by a water layer, capillary action forms a bond with the surface, further reducing friction performance and increasing adhesion [59].

2.5 Mechanism of Nanolubrication

Researchers have demonstrated the efficiency of nanolubrication to improve the tribological performance of tribological systems. Several advanced analytical techniques such as SEM and TEM help in the formation of different types of effective lubrication mechanisms by analyzing surfaces. There are four types of lubrication mechanisms [1].

(1) Rolling mechanism: Spherical nanoparticles exhibit such a mechanism. Then nanoparticles act like ball bearings and roll over the uneven surface. This type of mechanism is observed under low voltage conditions, as the shape of the nanoparticles deforms with increasing voltage.

(2) Formation of a protective layer: A chemical reaction between the nanoparticles and the surface forms a thin protective layer on the friction surface.

(3) Polishing mechanism: Nanoparticles reduce friction by smoothing surface irregularities during relative motion. This reduces shear forces between surfaces and improves tribological performance.

Table 2.4 Different mechanism of nanolubrication.

S.no	Lubricant	Type of nanoparticles	Mechanism	Size	References
1	SN90, SN150	WS2	Protective film	50-150 nm	[60]
2	Gear oil	Graphite	Ball bearing	55 nm	[61]
3	SAE20 W50	HbN	Protective film	50 nm	[62]
4	Synthetic oil, mineral oil, Vegetable oil	ZnO, CuO	Protective film	11.71nm - 4.35 nm	[63]
5	SAE30	CuO, TIO_2 Nano-diamond	Rolling	5 nm, 10 nm, 80 mm	[64]
6	PAO	Ni	Protective film	20 nm	[65]
7	Mineral oil	Cu	Mending	25 nm	[66]
8	Palm oil	TiO_2	Protective film		[67]
9	Supergear EP200	Carbon, Ag, Cu, Graphite, Fullerene	Rolling, Polishing	60 nm, 7 nm, 25 nm, less than 1 nm	[68]
10	Liquid Paraffin	SiO_2	Protective film	58-600 nm	[69]

Source: Adopted from Shafi WK, Charoo M. Nanolubrication systems: an overview. *Mater Today: Proc.* 2018;5(9):20621–20630.

(4) Repair mechanism: In this case, mass loss is compensated by the deposition of nanoparticles in the grooves of the surface.

Several nanolubrication mechanisms are summarized in Table 2.4.

2.6 Nanoparticle Properties Necessary for Nanolubrication

These important properties of nanoparticles for nanolubrication are as follows [1].

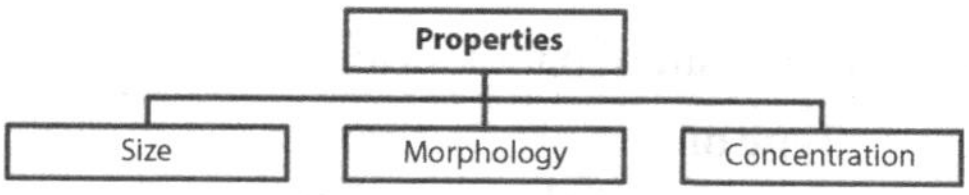

(1) Size: Studies have shown that an optimum size improves tribological performance. Nanoparticle size determines the following aspects in extreme nanolubrication:

(a) Control mechanical properties such as elastic hardness and physicochemical properties such as melting point. Hardness was found to be inversely proportional to particle size. As a result, a more durable surface will scratch a softer surface.
(b) Scale determines whether the nanoparticles can stay within the contact area. Larger particles tend to recede from the surface grooves.
(c) Nanoparticle size confirms dispersion stability [52, 70].

$$V_z = \frac{2(\rho_N - \rho_F) g r^2}{9\mu}$$

The Sed rate is Stokes' law given by where Vz is the decreasing velocity, r is the radius of the particle, and g is the acceleration due to gravity. From the equation, we can see that the deposition rate is directly proportional to the nanoparticle radius. The damping rate increases as the nanoparticle radius increases. As the descent rate increases, the nanoparticles are able to quiet the squeal, resulting in a lower settling rate. Peng *et al.* [69] studied the results of SiO_2 nanoparticles varying in size from 58 to 600 nm in liquid paraffin and found that nanoparticles with a size of 58 nm exhibited the most effective tribological performance in liquid lamp oil. Reeve *et al.* mixed HbN nanoparticles (70 nm, 0.5 μm, 1.5 μm, and 5 μm) into oil and found that minimal friction and wear were detected at 70 nm [71].

(2) Morphology: Particles of different shapes such as spherical, granular, nanosheet, nanotube and onion-like have different effects on tribological behavior. Spherical nanoparticles were found to have higher load-bearing capacity due to rolling results, whereas nanosheets deformed less than nanospheres and had less impact than nanospheres. Xu *et al.* [50] studied the morphological behavior of MoS_2 in oil and found that nanovesicles exhibited minimal COF and nanoplatelets exhibited maximum friction. Tebet *et al.* [72] studied C-like nanoparticles and found that their internal superimposed structure fixed parameters that were very important for their superiority in tribological performance.

(3) Concentration: The effectiveness of a nanolubricant is concentration-dependent. To get the best performance, an optimal concentration of nanolubricants is required. These optimal concentrations are system-dependent. So, different lubricants behave differently with a different system. Shaari *et al.* [73] mixed TiO_2 in oil at different concentrations. Oil was the most effective nanolubricant formulation [74]. SE 15W40 when mixed with HbN nanoparticles at an optimal concentration of 0.1 wt% gave the lowest friction and wear values, according to Luo *et al.* [75]. An optimal concentration of 0.1 wt % was reported when Al_2O_3 were mixed in lubricating oil at different concentrations of 0.05 wt%, 0.1 wt%, 0.5 wt%, and 1 wt%.

2.6.1 Nanolubricating Film Properties

In order to resist shear stresses and pressure, excellent nanolubricating films should have robust adhesive strength and cohesive strength. The most important properties of the film are its bonding characteristics and adherence to the surface. Either single molecular specie or mixed species should have a strong influence on the performance of the films. Due to physical and chemical assimilation, the molecules attach to the surface by adhesion and chemical bonding. The bonding force depends on the character of the bonds, the variety of bonds per molecule (in high relative molecular mass molecules), and therefore the molecular orientation and packing density per unit space. Surfaces may be of two types:

(1) Atomically flat surface with uniform surface: The energy throughout the surface is uniform. The molecular attachment to the surface is uniform and equally distributed.
(2) Atomically rough surfaces with non-uniform surface energy: The molecules tend to attach preferentially and form bonds with surface defect sites and high energy sites (steps, twins, dislocations, lattice defects, etc.).

The adhesive strength of the monolayer becomes harder because of surface reorganization and reconstruction and distribution as molecules. It is the result of the formation of islands and clusters by these molecules. Thus, the character of the surface and its surface energy square measure vital issues in adhesion and bonding of molecules on a surface [76].

2.6.2 Nanoparticles in Nanolubricants

Advancement in technology has driven many breakthroughs within the past few decades in terms of the invention of many forms of nanoparticles possessing application-specific properties. The accessible nanoparticles square measure are chiefly classified betting on their:

(1) morphology,
(2) shape,
(3) size, and
(4) properties [77].

A number of the foremost outstanding categories of nanoparticles are

(1) carbon-based nanoparticles,
(2) metal nanoparticles,
(3) tin alloy nanoparticles,
(4) ceramic nanoparticles,
(5) conductor-based nanoparticles,
(6) compound nanoparticles,
(7) composite nanoparticles, and
(8) lipid-based nanoparticles.

Among them, the most common nanoparticles are utilized as a dispersoid in nanolubricants square measure molybdenum disulfide (MoS_2), hexangular chemical element compound (h-BN), aluminum oxide (Al_2O_3), oxide (CuO), diamond, W disulfide (WS_2), philosopher's wool (ZnO), silver (Ag) and pigment (TiO_2) [78–83].

Aside from these, carbon-based nanoparticles have revolutionized the business owing to their inherent properties. The research project has discovered that the distinctive combination of physical and mechanical properties of carbon-based nanoparticles stems from hexangular arrays of sp2 hybridized carbon atoms. Specifically, the carbon-based nanoparticles comprise single and multi-walled carbon nanotubes, fullerenes, nanodiamonds, and graphene [77, 84].

2.7 Advantages of Nanolubricants

The Stribeck curve describes the friction levels of contacts with completely different film thickness-to-surface roughness ratios with lubrication regimes of the most important elements in automotive engines like piston rings, engine bearings, and valves. Combustion engine lubrication is classified into three main regimes:

(1) Boundary,
(2) mixed, and
(3) elasto-hydrodynamic lubrication [85].

In piston rings/liner assembly, the previous lubrication regimes are obtained over stroke betting on in-operation conditions. Generally, boundary lubrication exists underneath the result of low speed and high load conditions [86]. Hence, nanolubricant additives square measure is important in boundary lubrication owing to the upper friction constant [87]. Current challenges for reducing the friction associated with degreed wear need a filmable material for various operation conditions. Hence, most researchers have targeted the nanolubricant idea within the internal pair of combustion engines because the main strategy for suppressing the friction is constant, and therefore, the wear of contact surfaces, in a manner that may ultimately cause an improved tribological performance [88, 90].

Outlined below are the superior advantages of nanoparticle additives in comparison to conventional material additives [89]:

(1) Nanoparticles square measure are usually economical at temperature.
(2) Activation of nanoparticle surface
(3) Increase of extent and extreme little sizes
(4) Nanoparticles sized smaller than 100 nm have a thermal physical phenomenon above the fluids.
(5) Nanoparticles offer superior tribological performance as solid lubricating substances.

Nanoparticle-laden lubricants have an edge over microsized particles as they offer higher stability. The hazards and toxicity caused by additives to enhance the properties of the lubricants are largely eliminated by adding environmentally friendly, non-hazardous, and nontoxic nanoparticles. Lubricants' nanoparticles can reduce wear by bridging the gap between the tribo-pairs in operation.

Nanolubricants possess increased properties, particularly thermal and physical, that aid in extracting the generated heat, tends to excessive friction and wear [91–95]. The above advantage of nanolubricants makes them a strong contender and a strong candidate for application in many lubricating areas.

2.8 Nanoparticles Ability to Boost Grease Performance

Greases are described as solid to semi-liquid substances or as dispersions of thickeners in liquid lubricating substances [96]. Greases are classified by type of thickener (e.g., carboxylic acid soap), which accounts for between 100% of the grease [97]. These thickeners are fibrous, varying in length from 1 to 100 μm, the length being an order of magnitude greater than the diameter, and this size ratio determines the viscosity of the fat for a given thickener concentration related to consistency [98]. Fats, either mineral or artificial, make up the most important part of a fat formula. It is basically thickening that changes the operating temperature. The frictional performance of standard greases is primarily determined by the base oil and additives. In addition, frictional contact is full of currents generated by electric and magnetic fields [99]. A major challenge is also the torsional properties of the energy units. Unlike internal combustion engine (ICE) and electric vehicle (EV) motors produce more torsion and may provide maximum power in the low RPM range even at low speeds. The latter truth indicates the possibility of maximum boundary lubrication conditions. Moreover, once addressing higher motion speeds, which may reach up to 20,000 RPM [100], air could penetrate the parts and cause the fluids to foam, resulting in extreme stress on the motor. Therefore, low-viscosity greases are additional fitting for reducing torsion because the friction losses caused by rolling resistance will diminish significantly. It accelerates the decline in bearing life, thanks to the cutting in lubricating substance film and therefore, the realization of significant wear [97]. In such cases, nanolubricants will play a big role, thanks to their anti-vibration damping capability, let alone anti-wear (AW), anti-frictional (AF), and even extreme pressure (EP) superior performance.

The utilization of nanoparticles (NPs) in greases for electric vehicles (EVs) has been found to substantially influence their tribological performance. A groundbreaking transformation in mechanical engineering has occurred due to the innovative utilization of nanolubricants containing WS2 NPs featuring a closed-cage structure, often referred to as a fullerene-like IF. This pioneering work has been conducted by the academic team led by Professor Tenne [101], resulting in a paradigm shift in the understanding and application of lubrication in mechanical systems, particularly those associated with electric vehicles. However, most of the research in nanolubrication targets improving oil in homes in place of grease because the latter composition can be created in numerous formulations in homes [102], magnifying its evaluation complexity. However, the relatively little research on lubricating greases suggests that nanomaterials have numerous benefits that take place at the friction pair [103]. NPs can be categorized in keeping with their chemical nature into metallics, steel oxides, sulfides, clays, carbon-based and rare-earth as indicated in Table 2.5. A literature survey on the usage of several NP types as grease components is done, together with a significant discussion of the tribological action of these NPs.

Some studies claim that the round shape of NPs offers a good rolling mechanism, particularly when the NP diameter and film thickness are comparable [104–106]. Conversely, plate NPs have a propensity to slip [107–110].

The crystal structure can be highly entangled with lamellar grains. This is because weak Van der Waals forces connect adjacent layers and are easily slipped or destroyed by mere "stripping" of the layers [111]. One action of this is described by Sarno *et al.* This is because the dispersion of graphene compound NPs in oil provides friction reduction and wear resistance due to the fine and fragile layered NP structure, which reduces the shear stress and stops the interaction between metal interfaces. Earlier work by Dassenoy, emphasizing the degree, showed that metal disulfide NPs are more prone to exfoliate than to crystallize well. Moreover, under mixed or

Table 2.5 Nanoparticles in reducing wear scar diameter and friction coefficient in different grease bases.

Type of NPs	Size	Type of grease base	Reduction		Mechanism	References
			CoF	WSD		
TIO_2	21 nm	Calcium	72.3%	60%	Protective film	[115]
CNTs	D (10-12 nm).					
	L (1-20 um)					
ZnO	60 nm	Lithium	34.3%	8.3%	Sintering and mending	[116]
SigN.	100 mm					
G	10-20	Lithium	66%	91%	Protective film	[117]
MWCNTs	D (10-20 nm).					
	L (5-10 pm)					
G	NA	Lithium	15%	85%	Protective film	[118]
Cu	20-50 mm					
BN	NA	Lithium	46%	75%	Sliding (Ability to change adhesive wear to abrasive)	[119]
$Mc\&_2$	NA					
MoS	50-150	Lithium	42%	61%	Protective film	[120]
TIO_2	20 mm	Water-Glycerol	49.5%	97.8%	Rolling and mending	[121]
CeO_2	500 mm	Lithium	28%	13%	Polishing	[116]
rGO	500 mm	Lithium	60%	35%	Protective film	[122]
αAl_2O_3	40 mm	Lithium	Negative effect		Abrasion	
$CaCo_3$	50 mm	Lithium	33%	20%	Protective film	
$CaCo_3$	45 mm	Lithium	7.5%	12%	Protective film	[123]
CNTs	D (5 mm). L s mm)	Lithium	81.5%	63%	Protective film	[124]

(*Continued*)

Table 2.5 Nanoparticles in reducing wear scar diameter and friction coefficient in different grease bases. (*Continued*)

Type of NPs	Size	Type of grease base	Reduction		Mechanism	References
			CoF	WSD		
CaF_2	60 mm	Lithium	19%	29%	Protective film	[125]
G	NA	Lithium	50%	50%	Protective film	[126]
$Ca_{3},(BO_3)_2$	0 mm	Lithium	10%	33%	Tribochemical reaction and protective film	[127]
TIO_2,	35 mm	PTEF	14%	3%	Protective film	[128]
SiO_2	30 nm	PTEF	14%	Negative	Protective film	
SnO_2	20 mm	PTEF	Negative effect	effect	Deterioration of grease's mechanical properties	
NGT	2-6 pm	Lithium	No effect		No action	[129]
		Titanium	14.6-12.9%	2-2.3%	Tribochemical reaction and protective film	
TIO_2	40-60 nm	Lithium	40%	50%	Protective film	[130]
CuO	90-110 nm	Lithium	20%	60%	Protective film	
CuO	15-530 nm	Lithium	30%	13%	Polishing and mending	[131]
ZrO	12.7 mm	Lithium	48.8%	46.5%	Sintering	[132]
SiO_2	>100 mm	Lithium	26%	7%	Protective film	[133]
CS	>600 mm	Lithium	20%	40%	Rolling	[134]
Bi	10-30	Lithium	13.4%	19.2%	Mending	[135]
BN	255-955	Lithium	20%	10%	Protective film and sliding	[136]

Source: Adopted from Ahmed Abdalglil Mustafa, W, Dassenoy, F, Sarno, M, Senatore, A. A review on potentials and challenges of nanolubricants as promising lubricants for electric vehicles. *Lubrication Science*. 2022; 34(1): 1- 29.

boundary lubrication conditions, NPs can share some degree of compressive stress, resulting in the formation of strong self-laminating protective films or micropolishes and self-contacting friction surfaces [110, 112–114].

2.9 Tribological Performance of Nanolubricants

The tribological performance of nanolubricants depends on three parameters, namely:

(1) mechanical properties of a tribological system,
(2) lubrication, and
(3) chemical properties of the lubricating substance [137].

Energy Losses
As a result of friction, energy losses within the engine result in heating and damage to moving parts. A study reported that 75%–82% of total energy is lost within a vehicle, engine losses are between 68% and 72%, 12%–30% of fuel energy fails to move the vehicle [138], and the friction losses in vehicles are generally around 15% [139, 140]. Friction and wear together caused a total loss of energy of half an hour [141]. Numerous technologies, such as tire and bearing planning, tribology, and additives, can reduce friction-related energy losses. To reduce energy losses, the simplest lubricating substance should be used in order to beat the loss of energy. In addition, in order to minimize energy losses, using the simplest lubricant is important.

2.9.1 Mechanical Properties of a Tribological System

Figure 2.2 illustrates the slippy mechanism of the rubbing half and the relationship between material hardness and surface roughness in a tribological system. Lab tribometers are the only instruments that can measure these properties. Fabrics for checks typically have a Hardness Rockwell C (HRC) of 30–64 and surface roughness of .001 to 1.0 m.

To reduce and eliminate friction between two interacting components, lubrication is crucial. There are three recognized lubrication regimes, which are in accordance with the Stribeck curve:

(1) The first regime is boundary lubrication (BL).
(2) The second is mixed and elasto-hydrodynamic lubrication (EHL).
(3) The third is fluid mechanics lubrication (HL) [142].

Due to the unstable and thin tribofilm that forms on the rubbing surfaces, BL occurs when repeated loading/high load or low slippy speed generates heat and results in high wear and energy losses. As a result, lubricant additives are required to protect the friction surface(s) [143]. Shock loading is another factor that causes BL. Again, the lubrication regimes can change to a mixed/EHL regime and then to a fluid mechanics regime by lowering the load or increasing the slippy speed. As a result of improved fluid mechanics caused by the ability of a thicker tribofilm to separate the contacting surfaces, friction and wear can be reduced during this transition. A metric capacity unit can happen once a tribofilm separates two interacting surfaces. Due to the contacts being completely lubricated and the friction being controlled by viscous dragging forces, only fatigue occurs during this regime [143]. EHL is similar to the lubrication used in fluid mechanics. However, this lubrication is a part of the rolling motion. Mixed lubrication may have both metric and BL capacity units. The tribofilm had a film thickness that was around 1–3 times greater than surface roughness once below a mixed/EHL regime. The tribofilm had a film thickness that was three times greater than the surface roughness below the fluid mechanics regime [144]. Additionally, the tribofilm in EHL lubrication is much more diluted, resulting in greater pressure than in fluid mechanics lubrication. The constant of friction (COF) value of BL is often greater than 0.1, whereas the range of zero is covered by mixed/EHL lubrication. A 0.01 COF and metric capacity unit are included [145].

2.9.2 Physicochemical Properties of the Lubricant

For the best tribological performance, chemical properties of the lubricating substance is an important aspect. Differing kinds of oil have different properties, e.g., viscosity, density, and flash point, due to their origin from different feedstok.

Viscosity is the property of a liquid to resist to flow because of internal friction, and it is the most important and necessary property of the lubricating oil, and it depends on temperature and pressure, which ultimately decide tribological performance [146]. Higher viscosity indicates higher resistance to flow, leading to a thick tribofilm with lower potency. Moreover, higher viscosity leads to poor fuel atomization [147]. A lubricant with a high pressure-viscosity constant and high viscosity index will be the most common to ensure the tribofilm is stable under the conditions of elevated temperature and high loading.

The strength and stability of the tribofilm can be affected by additional physical features, such as pour purpose, flash and fireplace purpose, aerobic stability, and thermal stability [147]. Additionally, the tribochemical properties of lubricants with regard to purported surface active components show lubricity, load-carrying capacity, and wear resistance. They primarily serve as friction modifiers (FM), extreme pressure additives (EP), and anti-wear additives (AW) [137]. There are two main types of AW and EP additives: active and non-active. Through tribochemical reactions with the contact surface, active additives form a protective film to minimize wear. A non-active additive, on the other hand, forms a protective film when it transforms into a by-product, slowing down wear. As far as the application is concerned, AW and EP additives are analogous. In contrast, EP produces a thicker, more durable, and more suitable layer for high-speed operation, protecting films at a faster rate of reaction. Because FM provides a softer, torn-free protective film, it minimizes surface friction (sliding and rolling), thereby conserving energy. The COF of FM films is 0.01–0.05, which is lower than that of EP and AW films (0.06–0.15), but higher than that of EP and AW (0.001–0.009) in the hydrodynamic system [148]. These additives act primarily in BL and mixed/EHL lubrication systems to provide surface protection and lubricity enhancement [137].

2.10 Nanolubricants and Base Oils

2.10.1 Nanolubricants

Nanolubricants are nanoparticles/nanomaterials having particle sizes 1–100 nm used as material additive units. The use of nanolubricants in base oils significantly reduces friction and wear, resulting in exciting tribological properties. The main process is the physical or chemical synthesis of nanomaterials in dry form. This second process is to disperse the nanomaterials into the base oil by a mixing technique with or without dispersants or surfactants [149].

2.10.2 Base Oils

Three physical forms of lubricants are categorized:

(1) solid,
(2) semisolid, and
(3) liquid.

Lubricants are typically created from three different types of base oils. They are natural oil, synthetic oil, and biolubricant. The American Petroleum Institute (API) categorized base oil quality for various base oils according to API 1509, shown in Table 2.6 [150]. Group I to Group III base oils are made from refined petroleum. Group IV base oils are completely synthetic oils made from polyalphaolefins (PAO). Base oils in group V are those that are not in Groups I to IV. Various basic oils, such as siloxane, organophosphate, polyethylene glycol (PAG), polyester, and biolubricants, are included in this group.

By using fraction distillation, mineral oil is refined from crude oil. Oil is classified into three chemical compositions as follows:

(1) Paraffinic: Paraffinic contains predominantly paraffin straight-chain organic compounds.
(2) Naphthenic: Naphthenic hydrocarbons are in cyclic structures with no unsaturated bonds, and hydrocarbon also has a cyclic structure produced from benzol.
(3) Aromatic chemical compositions: Oil serves as the foundation of the economy and is a common lubricant used in trade, yet it is bad for both the environment and human health. They are typically used in bearings, gears, and engines, among other things. Artificial oil is made artificially from chemicals or hydrocarbons. Instead of using

Table 2.6 Base oil category as per its quality.

S.no	Base oil category	Made	Category
1	Group I (Solvent Refined)	Refined Petroleum	Mineral Oil
2	Group II (Hydrotreated)		
3	Group III (Hydrocracked)		
4	Group IV	Polyalphaolefins (PAO)	Synthetic Oil
5	Group IV (Those that are not in groups I to IV)	Siloxane,organophospate, polyalkyleneglycol (PAG), polyolester, and biolubricants etc.	----------

Source: Adopted from Jason, Y.J.J.; How, H.G.; Teoh, Y.H.; Chuah, H.G. A Study on the Tribological Performance of Nanolubricants. *Processes* 2020, 8, 1372.

full oil, this lubricating fluid will be manufactured in a facility using chemicals to change crude oil products. The creation of artificial oil has advantages over natural oils, including the capacity to lubricate under extreme cold or heat and improved wear protection [151, 152]. Artificial oils also offer numerous financial benefits, like lower maintenance and energy costs, and increased energy efficiency. Additionally, synthetic lubricants are made to meet the rigorous requirements of modern machinery [153]. On the other hand, some synthetic oils may harm the environment [154]. Oil is also appropriate for the manufacturing of biolubricants. However, vegetable oils including palm, coconut, rapeseed, and flower are specifically used to make biolubricants. Biolubricants can also be made using synthetic esters and crude oil products that meet biodegradability and toxicity requirements [146]. The first factor in determining a biolubricant's characteristics is its base stock. The characteristics represent chemistry, renewability, and degradability. Tribological characteristics are the most important requirement among these. Additionally, biolubricants are a supply because they have numerous sources, and are renewable, perishable, and environmentally beneficial. With their high lubricity, high body index, high flash point, low volatility, and great dispersion, biolubricants have many advantages over traditional oils [155, 156].

2.11 Various Types of Nanoparticles as Lubricant Additives

Nanoparticles contain metal, metal compounds, sulfides, nanocomposites, carbon nanoparticles, and complex compounds depending on their chemical makeup. According to Dai *et al.* [157], majority of the studies that were disseminated focused on metal-containing nanoparticles, which took up 72 % of the reviews. Studies that focused on carbon nanoparticles, nanocomposites, and rare earth compounds, only included 7%, 6%, and 7%, respectively.

The gilded NPs produced in this location have small particle sizes, high areas, low temperatures, and weak shear forces. As lubricant additives, they offer excellent tribological performance and self-repairing capabilities [158], along with atomic number 29, Bi, Sn, Fe, Ni, Al, Pd, Co, and Zn. The lubrication mechanisms of gilded NPs are as follows:

(1) The surface properties are modified, and the formation of a friction film separates the two friction surfaces, resulting in a promising friction performance.
(2) NP rolls between two friction surfaces, reducing friction and wear.
(3) The heat and pressure generated during manipulation densify the NPs on the damaged tracks, and this development is attributed to repair or sintering [157]. Padgurska *et al.* [159] studied the tribological properties of metal, Cu, and Co nanoparticles and the mixtures as lubricant additives to SAE-Ten oils. They reported that with nano Cu, single NPs reduce friction and wear individually or in combination, and the mixing of NPs is easier than pure NPs.

The tribological properties of Cu, Fe, and Zn NPs were studied comparatively among edible fats (rapeseed, soybean, canola, and olive), mineral oils, and artificial oils by Asadauskas *et al.* [160]. They reported that artificial oil, but not NPs, caused rock erosion, and with the addition of NPs, nano-Fe exhibited higher dispersion stability than nano-Cu and nano-Zn. Nano-Fe improved oil damage resistance and nano-Zn reduced oil wear and stirrup scratches. The addition of nanobismuth to light and heavy base oils reduced wear from 535 to 454 μm and 651 to 563 μm, and friction from 0.091 to 0.052 and 0.074 to 0.047 [161]. The addition of nano-Cu to carbon oil significantly reduced friction (23%) and wear (26%) [158]. An associate degree study on the tribological properties of nano-Ni in PAO_6 reported a 7%–30% reduction in wear and a 5%–45% reduction in friction [162]. Furthermore, the addition of nano-Al increased elasticity and improved friction and wear [163].

2.11.1 Metal Oxides

There are a variety of metal oxides, such as TiO_2, CuO, ZnO, Al_2O_3, Fe_3O_4, and $ZnAl_2O_4$ that are used as lubricant additives. As far as lubrication mechanism, sintering and repair result, and tribofilm formation are concerned, metal compound NPs are comparable to auriferous nanoparticles. The tribological behavior of ZnO and CuO in edible fat (sunflower and soybean), artificial oil, and oil was studied by Alves *et al.* [164]. They found that adding CuO to synthetic oils improved tribological properties and ZnO mineral-based lubricants exhibited superior friction and wear reduction, whereas adding NPs to cooking oils did not help in reduced wear. The addition of TiO_2 to water-based lubricants shows excellent tribological performance [165, 166]. On the other hand, SAE20W40 engine oil significantly reduced wear and reduced friction by a factor of 100 [167]. Luo *et al.* [168] used two types of tribometers to investigate the tribological properties of Al_2O_3 in pure oil. According to a four-ball tribometer, a typical reduction in friction constant (COF) is supposed to be 17.61% and a reduction in wear scar diameter (WSD) is 41.75%. Using compression rings, tribometer reduces friction by 23.92%. This can be achieved by forming a protective film on the contact surface and rewinding it. However, it should be noted that the addition of alumina adversely affects PAO and SAE75W-85 [169].

2.11.2 Metal Sulfides

Metal sulfides can be used in the form of solid or liquid additives, e.g., MoS_2, WS_2, FeS, and CuS, and have been used for many years. Nano-MoS_2 has been found to be better than micro-MoS_2 for reducing friction in liquid lubricants due to its smaller particle size [170]. The use of MoS_2 as a substance additive in dioctyl sebacate results in much lower friction and wear than micro-MoS_2 [171] due to the formation of an additional solid and demanding absorption film on the contact surface. Fullerene-like NPs (IFs) are a combination of layered compounds and hollow solid structures. Adding IF-MoS_2 and IF-WS_2 to PAO significantly improves the tribological properties [172]. Garzer *et al.* [173] investigated the tribological properties of chemically modified oils with the addition of MoS_2 and CuO, and nano-MoS_2 showed higher tribological properties than nano-CuO. A noteworthy study mentions the anti-friction ability of He nano-FeS with a particle size of 20–200 nm as an additive for motor oil substances [174]. The COF is significantly reduced by the addition of nano-FeS and exhibits persistent slipping behavior under dry and slippery conditions. A six samples indicates that the diffusion of S atoms to the friction surface creates a sulfur diffusion space and reduces friction.

2.11.3 Carbon-Based Nanoparticles

The use of carbon-based NPs as material additives continues to be relevant for nursing innovation. They embody diamond, graphene, and carbon. Diamond NPs in lamp oil have been reported to exhibit excellent tribological properties owing to the formation of a protective film separating friction surfaces [175]. It should be noted that the change in adhesion to wear mechanism improves the tribological behavior of this PAO at an optimal concentration of 0.2 wt% diamond NPs [176]. The fouling mechanisms of carbon NPs are the result of touch-up, tribofilm formation, and rolling. According to Bijwe *et al.* [177], dispersant-containing carbon gives him up to 80% best tribological performance. The improvement in AW properties is because of a mixture of tribofilming and touch-up results, and it is said that the mechanism of tribofilming contributes to the improvement in EP [178]. According to carbon compounds, material additives synthesized from waste carbon sources reduce frictional wear and surface roughness by up to 21.1%, 18.5%, and 42.3%, respectively. Graphene stands out in formal NPs due to its unique properties along with excellent mechanical, physical, and electrical properties. Moreover, graphene has been used in many applications and is commonly referred to as a supermaterial or all-in-one material in the world of materials science [179]. The addition of graphene to engine oils is said to increase friction and wear by up to 833 times, improve ball bearing performance, and provide graphene's superior strength properties [180]. A comparative study of modified natural flake carbon and modified graphene platelets in SN350 base oil was published by Sculpturer *et al.* Results indicated that the modified graphene platelets possessed higher tribological properties [181]. Graphene by exfoliation should also be mentioned [182, 183]. Ocladecylamine reduces graphene chemical nanolubricants in the frontier and hydrodynamic domains, reported by Vats *et al.* Results showed a 61.8% and 75% reduction in COF with these regimens, and a 92.5% significant remission of WSD within the border line regimen [184]. Moreover, the consistency of these nanolubricants improved over time through flow analysis.

2.11.4 Nanocomposites

These are multicomponent materials such as WC-Al_2O_3/graphene platelets, Cu/graphene compounds, TiO_2/SiO_2, Ag/graphene, carbon oxide/Cu, and Al_2O_3/TiO_2. Due to the synergistic effect of mixtures of NPs, nanocomposites may offer higher performance than individual NPs. It should be noted that graphene-containing nanocomposites exhibit excellent tribological performance [138]. Tests under 40N and 60N loads show 40.4% and 33.3% of the COF incorporated in the graphene lower than without graphene carried test. Moreover, adding graphene changes from severe abrasion to mild abrasion [185]. It is interesting to confirm the tribological properties of TiO_2/SiO_2 NPs in palm organic trimethlolpropane

(TMP) compounds, even without the use of surfactants. This is because of its stable dispersion stability as well as its effectiveness in reducing friction and wear. At the same time, the surface is increased as a result of repairing and sharpening [186]. Composites consisting of Cu-MoS_2 and Ag-MoS_2 reduce COF and generally improve wear resistance [187]. Copper/carbon fullerene nanocomposite reduces friction and wear by up to 23.7% and 33.5%, respectively [188]. Each study reported that the development of lubrication was because of the synergistic effect of nanocomposites.

2.11.5 Rare Earth Compounds

Rare earth compounds are used as lubricant additives or added to alternative NPs. Usually, the lubrication mechanism of rare earth element compounds is the formation of tribofilm or absorption film. La-doped Mg/Al superimposed double hydroxide NPs modified with dodecyl salt of sodium exhibits higher frictional properties than those not modified with ICE oil CD 15W-40. The lubrication mechanism ends when a tribofilm forms on the friction surface and reduces friction [138]. Hafnium compounds show excellent tribological properties even in greases containing metallic elements and lithium greases [138].

2.12 Recent Advancement in Nanolubrication

Water-based lubricants have grown in popularity recently. Due to their high thermal conductivity, water-based lubricants can be used as coolants during operations like milling, drilling, and other processes that call for cooling and appropriate lubrication. Water-based lubricants are an excellent substitute because they are highly fluid and environmentally friendly. Oil has been utilized as an addition over time to improve the lubricating characteristics of water, but it does not address the environmental problems associated with conventional grease-based lubricants. As a result, it is important to search for substitute additives. Recently, numerous studies indicated that introducing nanoparticles might be the answer. A lubricant's quality is increased by adding nanotubes, and environmental harm is reduced [189]. Lubricant industries are in dire need of innovative lubricants that, in addition to possessing good anti-wear (AW) and anti-friction (AF) qualities, can survive extreme pressure (EP) conditions. By adding nanoparticles (NPs) of various types and shapes, including spherical graphite (55 nm), lamellar multilayer graphene, talc (50–100 nm), spherical titanium (70 nm), and spherical MoDTC (20 nm), the overall lubrication potential of Polytetrafluoroethylene (PTFE)-based nano oils (oval-shaped, 200 nm) is intended to be improved. The weld loads were increased from 560 to 700 kg with a roughly 25% increase in performance when only 1% of MoDTC was used in combination with PTFE NPs. Further, the tribo-properties of oils can be modified by adding a special mix of NPs as additives [190].

2.13 Conclusion and Future Outlook

Nanolubricants nowadays are used as environmentally friendly lubricant additives. Nanoparticle additives are mostly utilized as anti-wear, anti-friction, and extreme pressure additives, and thus, they improve the overall performance of conventional lubricants such as metal working fluids, grease, and industrial oils. Their major properties, such as small enough size, thermal stability, a wide range of particle chemistries, and a rapid surface response rate without an induction phase, make them highly efficient for better lubrication. Their addition leads to better fuel efficiency, longer equipment operation times, and longer maintenance intervals leading to the reduction in power consumption, efficient lubrication, and better and longer machine life with higher mechanical efficiency. These nanolubricant additives improve tribological and rheological performance to a greater extent. Friction is always a major concern for all industries, automobiles, and vehicles. With increasing demand and the advancement of technology, there is high demand for production in rough and harsh environments.

Nanolubricant additives nowadays are proving to be highly efficient by both minimizing power consumption and reducing friction between two layers. The chemical composition of these additives plays an important role in improving the efficiency of the existing lubricants. Therefore, a thorough study and compatibility of these additives with the lubricant should be done before adding them as it will impact the frictional performance of instruments and machines. Not only the chemical composition of nanoadditives should be analyzed, a detailed analysis of existing oil must also be done as these oils act as a carrier for these additives. A proper combination of lubricant and nanoadditive, as well as concentration, should be selected for the specific design and conditions to get optimum results.

References

1. Shafi, W.K. and Charoo, M., Nanolubrication systems: An overview. *Mater. Today: Proc.*, 5, 9, 20621–20630, 2018.
2. Kumar, P. and Wani, M.F., Synthesis and tribological properties of graphene: A review. *J. Tribol.*, 13, 36–71, 2017.
3. Choi, S.U.S. and Eastman, J.A., *Proceedings of the ASME International Mechanical Engineering Congress and Exposition*, San Francisco, CA, pp. 12–47, 1995.
4. Eastman, J.A., Choi, S.U.S., Li, S., Yu, W., Thompson, L.J., Anomalously increased effective thermal conductivities of ethylene glycol-based nanofluids containing copper nanoparticles. *Appl. Phys. Lett.*, 78, 718–720, 2001.
5. Paul, G., Pal, T., Manna, I., Thermo-physical property measurement of nano-gold dispersed water based nanofluids prepared by chemical precipitation technique. *J. Colloid Interface Sci.*, 349, 434–7, 2010.
6. Paul, G., Sarkar, S., Pal, T., Das, P.K., Manna, I., Concentration and size dependence of nano-silver dispersed water based nanofluids. *J. Colloid Interface Sci.*, 371, 20–7, 2012.
7. Hernández Battez, A., Viesca, J.L., González, R., Blanco, D., Asedegbega, E., Osorio, A., Friction reduction properties of a CuO nanolubricant used as lubricant for a NiCrBSi coating. *Wear*, 268, 325–328, 2010.
8. Rashmi, W., Khalid, M., Xiao, Y.L., Gupta, T.C.S.M., Arwin, G.Z., Tribologicalstudies ongraphene/TMPbasednanolubricant. *J. Eng. Sci. Technol.*, 12, 365373, 2017, https://pureapps2.hw.ac.uk/ws/portalfiles/portal/16440665/12_2_6.pdf.
9. Nabil, M.F., Azmi, W.H., Hamid, K.A., Zawawi, N.N.M., Priyandoko, G., Mamat, R., Thermo-physical properties of hybrid nanofluids and hybrid nanolubricants: A comprehensive review on performance. *Int. Commun. Heat Mass Transf.*, 83, 30–39, 2017.
10. Dai, W., Kheireddin, B., Gao, H., Liang, H., Roles of nanoparticles in oil lubrication. *Tribol. Int.*, 102, 88–98, 2016.
11. Nomède-Martyr, N., Disa, E., Thomas, P., Romana, L., Mansot, J.L., Dubois, M., Guérin, K., Zhang, W., Hamwi, A., Effect of fluorination on the stability of carbon nanofibres in organic solventsInfluence de la fluoration sur la stabilité des nanofibres de carbone en dispersion dans des solvants organiques. *J. Fluor. Chem.*, 144, 10–16, 2012.
12. Bowden, E.P. and Tabor, D., *The Friction and Lubrication of Solids*, Oxford University Press, New York, 1950.
13. Gulzar, M., Masjuki, H., Kalam, M., Varman, M., Zulkifli, N., Mufti, R. *et al.*, Tribological performance of nanoparticles as lubricating oil additives. *J. Nanopart. Res.*, 18, 223, 2016.
14. Schneider, M.P., Plant-oil-based lubricants and hydraulic fluids. *J. Sci. Food Agric.*, 86, 1769–80, 2006.

15. Dhand, C., Dwivedi, N., Loh, X.J., Ying, A.N.J., Verma, N.K., Beuerman, R.W. *et al.*, Methods and strategies for the synthesis of diverse nanoparticles and their applications: A comprehensive overview. *RSC Adv.*, 5, 105003–37, 2015.
16. Schmidt, H., Nanoparticles by chemical synthesis, processing to materials and innovative applications. *Appl. Organomet. Chem.*, 15, 331–43, 2001.
17. Tran, T.H. and Nguyen, V.T., Copper oxide nanomaterials prepared by solution methods, some properties, and potential applications: A brief review. *Int. Sch. Res. Notices*, 2014, 1–14, 2014.
18. Thakkar, K.N., Mhatre, S.S., Parikh, R.Y., Biological synthesis of metallic nanoparticles. *Nanomed. Nanotechnol. Biol. Med.*, 6, 257–62, 2010.
19. Lam, C., Zhang, Y., Tang, Y., Lee, C., Bello, I., Lee, S., Large-scale synthesis of ultrafine Si nanoparticles by ball milling. *J. Cryst. Growth*, 220, 466–70, 2000.
20. Chin, P., Ding, J., Yi, J., Liu, B., Synthesis of FeS 2 and FeS nanoparticles by high-energy mechanical milling and mechanochemical processing. *J. Alloys Compd.*, 390, 255–60, 2005.
21. Xing, T., Sunarso, J., Yang, W., Yin, Y., Glushenkov, A.M., Li, L.H. *et al.*, Ball milling: A green mechanochemical approach for synthesis of nitrogen doped carbon nanoparticles. *Nanoscale*, 5, 7970–6, 2013.
22. De Carvalho, J., de Medeiros, S., Morales, M., Dantas, A., Carriço, A., Synthesis of magnetite nanoparticles by high energy ball milling. *Appl. Surf. Sci.*, 275, 84–7, 2013.
23. Pérez-Tijerina, E., Pinilla, M.G., Mejía-Rosales, S., Ortiz-Méndez, U., Torres, A., José-Yacamán, M., Highly size-controlled synthesis of Au/Pd nanoparticles by inert-gas condensation. *Faraday Discuss.*, 138, 353–62, 2008.
24. Ward, M., Brydson, R., Cochrane, R., Mn nanoparticles produced by inert gas condensation. *J. Phys. Conf. Ser. IOP Publishing*, 26, 1, 296–299, 2006.
25. Cross, C., Hemminger, J., Penner, R., Physical vapor deposition of one-dimensional nanoparticle arrays on graphite: seeding the electrodeposition of gold nanowires. *Langmuir*, 23, 10372–9, 2007.
26. Gondal, M., Drmosh, Q., Yamani, Z., Saleh, T., Synthesis of ZnO 2 nanoparticles by laser ablation in liquid and their annealing transformation into ZnO nanoparticles. *Appl. Surf. Sci.*, 256, 298–304, 2009.
27. König, D., Richter, K., Siegel, A., Mudring, A.V., Ludwig, A., High-throughput fabrication of Au–Cu nanoparticle libraries by combinatorial sputtering in ionic liquids. *Adv. Funct. Mater.*, 24, 2049–56, 2014.
28. Dumitrache, F., Morjan, I., Alexandrescu, R., Ciupina, V., Prodan, G., Voicu, I. *et al.*, Iron–iron oxide core–shell nanoparticles synthesized by laser pyrolysis followed by superficial oxidation. *Appl. Surf. Sci.*, 247, 25–31, 2005.
29. Alexandrescu, R., Morjan, I., Scarisoreanu, M., Birjega, R., Popovici, E., Soare, I. *et al.*, Structural investigations on TiO 2 and Fe-doped TiO 2 nanoparticles synthesized by laser pyrolysis. *Thin Solid Films*, 515, 8438–45, 2007.

30. Nemade, K. and Waghuley, S., Synthesis of MgO nanoparticles by solvent mixed spray pyrolysis technique for optical investigation. *Int. J. Met.*, 2014, 1–4, 2014.
31. Behbahani, A., Rowshanzamir, S., Esmaeilifar, A., Hydrothermal synthesis of zirconia nanoparticles from commercial zirconia. *Proc. Eng.*, 42, 908–17, 2012.
32. Liu, B. and Zeng, H.C., Hydrothermal synthesis of ZnO nanorods in the diameter regime of 50 nm. *J. Am. Chem. Soc*, 125, 4430–1, 2003.
33. Beach, E.R., Shqau, K., Brown, S.E., Rozeveld, S.J., Morris, P.A., Solvothermal synthesis of crystalline nickel oxide nanoparticles. *Mater. Chem. Phys.*, 115, 371–7, 2009.
34. Srivastava, V., Gusain, D., Sharma, Y.C., Synthesis, characterization and application of zinc oxide nanoparticles (n-ZnO). *Ceram. Int.*, 39, 9803–8, 2013.
35. Karuppuchamy, S. and Jeong, J.M., Synthesis of nano-particles of TiO2 by simple aqueous route. *J. Oleo Sci.*, 55, 263–6, 2006.
36. Lim, H., Ahmad, A., Hamzah, H., Synthesis of zirconium oxide nanoparticle by sol-gel technique. *AIP Conf. Proc. AIP*, 1571, 812–6, 2013.
37. Oseghe, E.O., Ndungu, P.G., Jonnalagadda, S.B., Synthesis, characterization, and application of TiO2 nanoparticles–effect of pH adjusted solvent. *J. Adv. Oxid. Technol.*, 18, 253–63, 2015.
38. Sun, Y., Mayers, B., Herricks, T., Xia, Y., Polyol synthesis of uniform silver nanowires: A plausible growth mechanism and the supporting evidence. *Nano Lett.*, 3, 955–60, 2003.
39. Wang, Y., Choi, S.I., Zhao, X., Xie, S., Peng, H.C., Chi, M. *et al.*, Polyol synthesis of ultrathin Pd nanowires via attachment-based growth and their enhanced activity towards formic acid oxidation. *Adv. Funct. Mater.*, 24, 131–9, 2014.
40. Song, J.Y., Jang, H.-K., Kim, B.S., Biological synthesis of gold nanoparticles using Magnolia kobus and Diopyros kaki leaf extracts. *Process Biochem.*, 44, 1133–8, 2009.
41. Song, J.Y. and Kim, B.S., Rapid biological synthesis of silver nanoparticles using plant leaf extracts. *Bioprocess Biosyst. Eng.*, 32, 79, 2009.
42. Vigneshwaran, N., Ashtaputre, N., Varadarajan, P., Nachane, R., Paralikar, K., Balasubramanya, R., Biological synthesis of silver nanoparticles using the fungus Aspergillus flavus. *Mater. Lett.*, 61, 1413–8, 2007.
43. He, S., Guo, Z., Zhang, Y., Zhang, S., Wang, J., Gu, N., Biosynthesis of gold nanoparticles using the bacteria Rhodopseudomonas Capsulata. *Mater. Lett.*, 61, 3984–7, 2007.
44. Padil, V.V.T. and Černík, M., Green synthesis of copper oxide nanoparticles using gum karaya as a biotemplate and their antibacterial application. *Int. J. Nanomedicine*, 8, 889, 2013.

45. Mukherjee, P., Senapati, S., Mandal, D., Ahmad, A., Khan, M.I., Kumar, R. *et al.*, Extracellular synthesis of gold nanoparticles by the fungus Fusarium oxysporum. *ChemBioChem*, 3, 461–3, 2002.
46. Raudabaugh, D.B., Tzolov, M.B., Calabrese, J.P., Overton, B.E., Synthesis of silver nanoparticles by a Bryophilous Rhizoctonia species. *Nanomater. Nanotechnol.*, 3, 2, 2013.
47. Yu, W. and Xie, H., A review on nanofluids: Preparation, stability mechanisms, and applications. *J. Nanomater.*, 2012, 1, 2012.
48. Battez, A.H., Rico, J.F., Arias, A.N., Rodriguez, J.V., Rodriguez, R.C., Fernandez, J.D., The tribological behaviour of ZnO nanoparticles as an additive to PAO6. *Wear*, 261, 256–63, 2006.
49. Koshy, C.P., Rajendrakumar, P.K., Thottackkad, M.V., Analysis of tribological and thermo-physical properties of surfactant-modified vegetable oil-based CuO nano-lubricants at elevated temperatures-an experimental study. *Tribol. Online*, 10, 344–53, 2015.
50. Xu, Z.Y., Hu, K.H., Han, C.L., Hu, X.G., Xu, Y.F., Morphological influence of molybdenum disulfide on the tribological properties of rapeseed oil. *Tribol. Lett.*, 49, 513–24, 2013.
51. Bönnemann, H., Botha, S., Bladergroen, B., Linkov, V., Monodisperse copperand silver-nanocolloids suitable for heat-conductive fluids. *Appl. Organomet. Chem.*, 19, 768–73, 2005.
52. Akbulut, M., Nanoparticle-based lubrication systems. *J. Powder Metall. Min*, 1, e101, 2012.
53. Guo, D., Xie, G., Luo, J., Mechanical properties of nanoparticles: basics and applications. *J. Phys. D Appl. Phys.*, 47, 013001, 2013.
54. Keesom, W., On the deduction of the equation of state from Boltzmann's entropy principle'. *KNAW, Proceedings*, pp. 240–56, 1912.
55. Debye, P., Molecular forces and their electrical interpretation. *PhysikalischeZeitschrift*, 22, 302–8, 1921.
56. London, F., The general theory of molecular forces. *Trans. Faraday Soc.*, 33, 8b–26, 1937.
57. Israelachvili, J.N., *Intermolecular and Surface Forces*, Academic Press, 2011, https://doi.org/10.1016/C2009-0-21560-1.
58. Quon, R., Ulman, A., Vanderlick, T., Impact of humidity on adhesion between rough surfaces. *Langmuir.*, 16, 8912–6, 2000.
59. Israelachvili, J., Maeda, N., Rosenberg, K.J., Akbulut, M., Effects of sub-Ångstrom (pico-scale) structure of surfaces on adhesion, friction, and bulk mechanical properties. *J. Mater. Res.*, 20, 1952–72, 2005.
60. Greenberg, R., Halperin, G., Etsion, I., Tenne, R., The effect of WS 2 nanoparticles on friction reduction in various lubrication regimes. *Tribol. Lett.*, 17, 179–86, 2004.
61. Lee, C.-G., Hwang, Y.-J., Choi, Y.-M., Lee, J.-K., Choi, C., Oh, J.-M., A study on the tribological characteristics of graphite nano lubricants. *Int. J. Precis. Eng. Manuf.*, 10, 85–90, 2009.

62. Charoo, M. and Wani, M.F., Tribological properties of h-BN nanoparticles as lubricant additive on cylinder liner and piston ring. *Lubr. Sci.*, 29, 241–54, 2017.
63. Alves, S., Barros, B., Trajano, M., Ribeiro, K., Moura, E., Tribological behavior of vegetable oil-based lubricants with nanoparticles of oxides in boundary lubrication conditions. *Tribol. Int.*, 65, 28–36, 2013.
64. Wu, Y., Tsui, W., Liu, T., Experimental analysis of tribological properties of lubricating oils with nanoparticle additives. *Wear*, 262, 819–25, 2007. Shafi, W.K. *et al.*, *Mater. Today Proc.*, 5, 20621–20630, 2018.
65. Chou, R., Battez, A.H., Cabello, J., Viesca, J., Osorio, A., Sagastume, A., Tribological behavior of polyalphaolefin with the addition of nickel nanoparticles. *Tribol. Int.*, 43, 2327–32, 2010.
66. Choi, Y., Lee, C., Hwang, Y., Park, M., Lee, J., Choi, C. *et al.*, Tribological behavior of copper nanoparticles as additives in oil. *Curr. Appl. Phys.*, 9, e124–e7, 2009.
67. Zulkifli, N., Kalam, M., Masjuki, H., Yunus, R., Experimental analysis of tribological properties of biolubricant with nanoparticle additive. *Proc. Eng.*, 68, 152–7, 2013.
68. Choi, C., Jung, M., Choi, Y., Lee, J., Oh, J., Tribological properties of lubricating oilbased nanofluids with metal/carbon nanoparticles. *J. Nanosci. Nanotechnol.*, 11, 368–71, 2011.
69. Peng, D.-X., Chen, C.-H., Kang, Y., Chang, Y.-P., Chang, S.-Y., Size effects of SiO2 nanoparticles as oil additives on tribology of lubricant. *Ind. Lubr. Tribol.*, 62, 111–20, 2010.
70. Min, Y., Akbulut, M., Kristiansen, K., Golan, Y., Israelachvili, J., The role of interparticle and external forces in nanoparticle assembly. *Nat. Mater.*, 7, 527–38, 2008.
71. Reeves, C.J., Menezes, P.L., Lovell, M.R., Jen, T.-C., The size effect of boron nitride particles on the tribological performance of biolubricants for energy conservation and sustainability. *Tribol. Lett.*, 51, 437–52, 2013.
72. Tevet, O., Von-Huth, P., Popovitz-Biro, R., Rosentsveig, R., Wagner, H.D., Tenne, R., Friction mechanism of individual multilayered nanoparticles. *PNAS*, 108, 19901–6, 2011.
73. Shaari, M.Z., Roselina, N.N., Kasolang, S., Hyie, K.M., Murad, M.C., Bakar, M.A.A., Investigation of tribological properties of palm oil biolubricant modified nanoparticles. *J. Teknol.*, 76, 69–73, 2015.
74. Wan, Q., Jin, Y., Sun, P., Ding, Y., Tribological behaviour of a lubricant oil containing boron nitride nanoparticles. *Proc. Eng.*, 102, 1038–45, 2015.
75. Luo, T., Wei, X., Huang, X., Huang, L., Yang, F., Tribological properties of Al 2 O 3 nanoparticles as lubricating oil additives. *Ceram. Int.*, 40, 7143–9, 2014.
76. Hsu, S.M., Nano-lubrication: Concept and design. *Tribol. Int.*, 37, 7, 537–545, 2004.

77. Khan, I., Saeed, K., Khan, I., Nanoparticles: Properties, applications and toxicities. *Arab. J. Chem.*, 12, 7, 908–931, 2019, https://doi.org/10.1016/j.arabjc.2017.05.011.
78. Hernández Battez, A., Viesca, J.L., González, R., Blanco, D., Asedegbega, E., Osorio, A., Friction reduction properties of a CuO nanolubricant used as lubricant for a NiCrBSi coating. *Wear*, 268, 325–328, 2010.
79. Yu, W., France, D.M., Choi, S.U.S., Routbort, J.L., Review and assessment of nanofluid technology for transportation and other applications. pp. 6–75, Argonne National Lab. (ANL), Argonne, IL (United States), 2007, https://doi.org/10.2172/919327.
80. Ben Difallah, B., Kharrat, M., Dammak, M., Monteil, G., Improvement in the tribological performance of polycarbonate via the incorporation of molybdenum disulfide particles. *Tribol. Trans.*, 57, 806–813, 2014.
81. Mosleh, M., Atnafu, N.D., Belk, J.H., Nobles, O.M., Modification of sheet metal forming fluids with dispersed nanoparticles for improved lubrication. *Wear*, 267, 1220–1225, 2009.
82. Gara, L. and Zou, Q., Friction and wear characteristics of water-based. ZnO and Al_2O_3 nanofluids. *Tribol. Trans.*, 55, 345–350, 2012.
83. Ghaednia, H., Hossain, M.S., Jackson, R.L., Tribological performance of silver nanoparticle–enhanced polyethylene glycol lubricants. *Tribol. Trans.*, 59, 585–592, 2016.
84. Cha, C., Shin, S.R., Annabi, N., Dokmeci, M.R., Khademhosseini, A., Carbon-based nanomaterials: Multifunctional materials for biomedical engineering. *ACS Nano*, 7, 2891–2897, 2013.
85. Ali, M.K.A., Xianjun, H., Mai, L. *et al.*, Improving the tribological characteristics of piston ring assembly in automotive engines using Al_2O_3 and TiO_2 nanomaterials as nano-lubricant additives. *Tribol. Int.*, 103, 540–54, 2016.
86. Ali, M.K.A., Xianjun, H., Turkson, R.F. *et al.*, An analytical study of tribological parameters between piston ring and cylinder liner in internal combustion engines. *Proc. Inst. Mech. Engineers K J. Multi-body Dyn.*, 230, 4, 329–349, 2016.
87. Ali, M.K.A. and Xianjun, H., Improving the tribological behavior of internal combustion engines via the addition of nanoparticles to engine oils. *Nanotechnol. Rev.*, 4, 4, 347–358, 2015.
88. Ali, M.K.A., Xianjun, H., Elagouz, A. *et al.*, Minimizing of the boundary friction coefficient in automotive engines using Al_2O_3 and TiO_2 nanoparticles. *J. Nanopart. Res.*, 18, 377, 2016.
89. Martin, J.M. and Ohmae, N., Nanolubricants, colloidal lubrication: General principles, in: *Nanolubricants*, pp. 1–3, John Wiley, Sons, England, 2008.
90. Ali, M.K.A., Xianjun, H., Mai, L. *et al.*, Reducing frictional power losses and improving the scuffing resistance in automotive engines using hybrid nanomaterials as nano-lubricant additives. *Wear.*, 364-365, 270–281, 2016.

91. Rasheed, A.K., Khalid, M., Rashmi, W., Gupta, T.C.S.M., Chan, A., Graphene based nanofluids and nanolubricants – Review of recent developments. *Renew. Sustain. Energy Rev.*, 63, 346–362, 2016.
92. Rasheed, A.K., Khalid, M., Javeed, A., Rashmi, W., Gupta, T.C.S.M., Chan, A., Heat transfer and tribological performance of graphene nanolubricant in an internal combustion engine. *Tribol. Int.*, 103, 504–515, 2016.
93. Ali, M.K.A., Xianjun, H., Abdelkareem, M.A.A., Gulzar, M., Elsheikh, A.H., Investigation of the tribological behavior of mineral lubricant using copper oxide nano additives. *Tribol. Int.*, 124, 209–229, 2018.
94. Zin, V., Barison, S., Agresti, F., Colla, L., Pagura, C., Fabrizio, M., Improved tribological and thermal properties of lubricants by graphene based nano-additives. *RSC Adv.*, 6, 59477–59486, 2016.
95. Chandrabhan, S.R., Jayan, V., Parihar, S.S., Ramaprabhu, S., Development of a nitrogen-doped 2D material for tribological applications in the boundary-lubrication regime. *Beilstein J. Nanotechnol.*, 8, 1476–1483, 2017.
96. *Annual Book of ASTM Standards. ASTM D 288*, ASTM, ASTM International, 1961.
97. Lugt, P.M., Modern advancements in lubricating grease technology. *Tribol Int.*, 97, 467–477, 2016. https://doi.org/10.1016/j. triboint.2016.01.045.
98. Scarlett, N.A., Use of grease in rolling bearings. *Proc. Inst. Mech. Eng. Conf. Proc.*, 182, 1, 585–624, 1967. https://doi.org/10.1243/ PIME_CONF_ 1967_182_045_02.
99. Preisinger, G., *Cause and effect of bearing currents in frequency converter driven electrical motors: investigations of electrical properties of rolling bearings*, TU Wien Academic Press, 2002. http://hdl.handle.net/20.500.12708/186295
100. Farfan-Cabrera, L.I., Tribology of electric vehicles: A review of critical components, current state and future improvement trends. *Tribol Int.*, 138, 473486, 2019. https://doi.org/10.1016/j. triboint.2019.06.029.
101. Rapoport, L., Fleischer, N., Tenne, R., Fullerene-like WS2 nanoparticles: Superior lubricants for harsh conditions. *Adv. Mater.*, 15, 78, 651–655, 2003. https:// doi.org/10.1002/adma. 200301640.
102. Vengudusamy, B., Enekes, C., Spallek, R., On the film forming and friction behaviour of greases in rolling/sliding contacts. *Tribol Int.*, 129, 323337, 2019. https://doi.org/10.1016/j.triboint.2018.08.026.
103. Qiang, H., Wang, T., Qu, H., Zhang, Y., Li, A., Kong, L., Tribological and rheological properties of nanorods–Al2O3 as additives in grease. *Proc. Inst. Mech. Eng. J. J. Eng. Tribol.*, 233, 4, 605–614, 2019. https://doi.org/10. 1177/1350650118787403.
104. Azman, N.F., Samion, S., Sot, M.N.H.M., Investigation of tribological properties of CuO/palm oil nanolubricant using pin-on disc tribometer. *Green Mater.*, 6, 1, 30–37, 2018. https://doi. org/10.1680/jgrma.17.00026.
105. Lahouij, I., Dassenoy, F., Vacher, B., Martin, J.-M., Real time TEM imaging of compression and shear of single fullerene-like MoS2 nanoparticle. *Tribol. Lett.*, 45, 1, 131–141, 2012. https://doi.org/10.1007/s11249-011-9873-8.

106. Chinas-Castillo, F. and Spikes, H.A., Mechanism of action of colloidal solid dispersions. *J. Tribol.*, 125, 3, 552–557, 2003. https://doi. org/10.1115/1.1537752.
107. He, Q., Li, A., Guo, Y., Liu, S., Zhang, Y., Kong, L., Tribological properties of nanometer cerium oxide as additives in lithium grease. *J. Rare Earths*, 36, 2, 209214, 2018. https://doi.org/10. 1016/j.jre.2017.09.004.
108. Xie, H., Jiang, B., He, J., Xia, X., Pan, F., Lubrication performance of MoS2 and SiO2 nanoparticles as lubricant additives in magnesium alloy-steel contacts. *Tribol. Int.*, 93, 63–70, 2016. https:// doi.org/10.1016/j.triboint.2015.08.009.
109. Taha-Tijerina, J., Peña-Paras, L., Narayanan, T.N. *et al.*, Multifunctional nanofluids with 2D nanosheets for thermal and tribological management. *Wear*, 302, 1–2, 1241–1248, 2013. https://doi.org/10.1016/j.wear.2012.12.010.
110. Hu, K.H., Liu, M., Wang, Q.J., Xu, Y.F., Schraube, S., Hu, X.G., Tribological properties of molybdenum disulfide nanosheets by monolayer restacking process as additive in liquid paraffin. *Tribol. Int.*, 42, 1, 33–39, 2009. https:// doi. org/10.1016/j.triboint. 2008.05.016.
111. Yan, C., Zeng, Q., Hao, Y., Xu, Y., Zhou, M., Friction-induced hardening behaviors and tribological properties of 60NiTi alloy lubricated by lithium grease containing nano-BN and MoS 2. *Tribol. Trans.*, 62, 5, 812–820, 2019. https://doi.org/10.1080/ 10402004.2019.1619889.
112. Kato, H. and Komai, K., Tribofilm formation and mild wear by tribo-sintering of nanometer-sized oxide particles on rubbing steel surfaces. *Wear*, 262, 1–2, 36–41, 2007. https://doi.org/10. 1016/j.wear.2006.03.046.
113. Zulkifli, N.W.M., Kalam, M.A., Masjuki, H.H., Yunus, R., Experimental analysis of tribological properties of biolubricant with nanoparticle additive. *Proc. Eng.*, 68, 152–157, 2013. https://doi. org/10.1016/j.proeng.2013.12.161.
114. Ahmed Abdalglil Mustafa, W., Dassenoy, F., Sarno, M., Senatore, A., A review on potentials and challenges of nanolubricants as promising lubricants for electric vehicles. *Lubr. Sci.*, 34, 1, 1–29, 2022.
115. Mohamed, A., Ali, S., Osman, T.A., Kamel, B.M., Development and manufacturing an automated lubrication machine test for nano grease. *J. Mater. Res. Technol.*, 9, 2, 2054–2062, 2020. https://doi.org/10.1016/j.jmrt.2019.12.038.
116. He, Q., Li, A., Wang, Z., Zhang, Y., Kong, L., Yang, K., Tribological behavior of ZnO-Si 3 N 4 nanoparticles-based lubricating grease. *J. Exp. Nanosci.*, 13, 1, 231–244, 2018. https://doi.org/10. 1080/17458080.2018.1511923.
117. Ashour, M.E., Osman, T.A., Khattab, A., Elshalakny, A.B., Novel tribological behavior of hybrid MWCNTs/MLNGPs as an additive on Lithium grease. *J. Tribol.*, 139, 4, 1–7, 2017. https://doi. org/10.1115/1.4035345.
118. Jing, W., Xiaochuan, G., Yan, H.M.J., Weigui, Z., Synthesis and tribological properties of graphene-copper nanoparticle composites as lithium grease additive. *China Pet. Process. Petrochem. Technol.*, 19, 4, 113–122, 2017.
119. Yan, C., Zeng, Q., Hao, Y., Xu, Y., Zhou, M., Friction-induced hardening behaviors and tribological properties of 60NiTi alloy lubricated by lithium grease containing nano-BN and MoS 2. *Tribol. Trans.*, 62, 5, 812–820, 2019. https://doi.org/10.1080/ 10402004.2019.1619889.

120. Rawat, S.S., Harsha, A.P., Agarwal, D.P., Kumari, S., Khatri, O.P., Pristine and alkylated MoS2 nanosheets for enhancement of tribological performance of paraffin grease under boundary lubrication regime. *J. Tribol.*, 141, 7, 112, 2019. https://doi.org/ 10.1115/1.4043606.
121. Wu, H., Zhao, J., Xia, W. *et al.*, A study of the tribological behaviour of TiO 2 nano-additive water-based lubricants. *Tribol Int.*, 109, 398–408, 2017. https:// doi.org/10.1016/j.triboint.2017.01.013.
122. Singh, J., Kumar, D., Tandon, N., Tribological and vibration studies on newly developed nanocomposite greases under boundary lubrication regime. *J. Tribol.*, 140, 3, 1–10, 2018. https://doi. org/10.1115/1.4038100.
123. Ji, X., Chen, Y., Zhao, G., Wang, X., Liu, W., Tribological properties of CaCO3 nanoparticles as an additive in lithium grease. *Tribol. Lett.*, 41, 1, 113119, 2011. https://doi.org/10.1007/s11249-010-9688-z.
124. Mohamed, A., Osman, T.A., Khattab, A., Zaki, M., Tribological behavior of carbon nanotubes as an additive on Lithium grease. *J. Tribol.*, 137, 1, 1–5, 2015. https://doi.org/10.1115/1.4028225.
125. Wang, L., Wang, B., Wang, X., Liu, W., Tribological investigation of CaF2 nanocrystals as grease additives. *Tribol. Int.*, 40, 7, 1179–1185, 2007. https:// doi. org/10.1016/j.triboint.2006.12.003.
126. Cheng, Z.-L. and Qin, X.-X., Study on friction performance of graphene-based semi-solid grease. *Chin. Chem. Lett.*, 25, 9, 1305–1307, 2014. https:// doi.org/ 10.1016/j.cclet.2014.03.010.
127. Zhao, G., Zhao, Q., Li, W., Wang, X., Liu, W., Tribological properties of nano-calcium borate as lithium grease additive. *Lubr. Sci.*, 26, 1, 43–53, 2014. https:// doi.org/10.1002/ls.1227.
128. Ge, X., Xia, Y., Cao, Z., Tribological properties and insulation effect of nanometer TiO2 and nanometer SiO2 as additives in grease. *Tribol. Int.*, 92, 454461, 2015. https://doi.org/10.1016/j. triboint.2015.07.031.
129. Niu, M. and Qu, J., Tribological properties of nano-graphite as an additive in mixed oil based titanium complex grease. *RSC Adv.*, 8, 73, 133–144, 2018. https://doi.org/ 10.1039/C8RA08109C.
130. Chang, H., Lan, C.-W., Chen, C.-H., Kao, M.-J., Guo, J.-B., Anti-wear and friction properties of nanoparticles as additives in the lithium grease. *Int. J. Precis. Eng. Manuf.*, 15, 10, 2059–2063, 2014. https://doi.org/10.1007/ s12541-014-0563-y.
131. Zheng, B., Zhou, J., Jia, X., He, Q., Friction and wear property of lithium grease contained with copper oxide nanoparticles. *Appl. Nanosci.*, 10, 4, 13551367, 2020. https://doi.org/10.1007/ s13204-019-01219-7 233.
132. Xia, X.T., Zhang, Y.P., Zhang, Y.Z., Chen, S.C., Influence of ZrO2 nanoparticle as additive on tribological property of Lithium grease. *Appl. Mech. Mater.*, 26, 83–87, 2010.
133. He, Q., Li, A., Guo, Y., Liu, S., Kong, L.-H., Effect of nanometer silicon dioxide on the frictional behavior of lubricating grease. *Nanomater. Nanotechnol.*, 7, 593, 2017. https://doi.org/10.1177/ 1847980417725933.

134. Li, T., Pan, B., Wang, C., Zhang, S., Huang, S., Effect of eco-friendly carbon microspheres on the tribological behavior of lithium lubricating grease. *UPB Sci. Bull. Ser. B.*, 81, 57–68, 2019.
135. Zhang, X., Zhang, Z., Zhao, F., Qiu, T., Preparation and tribological properties of bi nanoparticles as a lithium grease additive. *Optoelectron. Adv. Mater. Rapid Commun.*, 8, 500–505, 2014.
136. Wang, T., Li, Z., Li, J., He, Q., Impact of boron nitride nanoparticles on the wear property of lithium base grease. *J. Mater. Eng. Perform.*, 29, 8, 4991–5000, 2020. https://doi.org/10. 1007/s11665-020-05008-0.
137. Chan, C.-H., Tang, S.W., Mohd, N.K., Lim, W.H., Yeong, S.K., Idris, Z., Tribological behavior of biolubricant base stocks and additives. *Renew. Sustain. Energy Rev.*, 93, 145–157, 2018.
138. Jason, Y.J.J., How, H.G., Teoh, Y.H., Chuah, H.G.A., Study on the tribological performance of nanolubricants. *Processes*, 8, 1372, 2020.
139. Jordan, H., Kailyn, S., Jason, D., *Energy Education—Energy Loss*, Available online: https://energyeducation.ca/encyclopedia/Energy_loss (accessed on 20 February 2020).
140. Government, U.S., *Where the Energy Goes: Gasoline Vehicles*, Available online: https://www.fueleconomy. gov/feg/atv.shtml (accessed on 20 February 2020).
141. Shafi, W.K., Raina, A., Haq, M.I.U., Friction and wear characteristics of vegetable oils using nanoparticles for sustainable lubrication. *Tribol. Mater. Surf. Interfaces*, 12, 27–43, 2018.
142. Kalin, M., Velkavrh, I., Vižintin, J., The stribeck curve and lubrication design for non-fully wetted surfaces. *Wear*, 267, 1232–1240, 2009.
143. Hamrock, B.J., Schmid, S.R., Jacobson, B.O., *Fundamentals of Fluid Film Lubrication*, CRC Press, Boca Raton, FL, USA, 2004.
144. Maru, M.M. and Tanaka, D.K., Consideration of stribeck diagram parameters in the investigation on wear and friction behavior in lubricated sliding. *J. Braz. Soc Mech. Sci. Eng.*, 29, 55–62, 2007.
145. Bruce, R.W., *Handbook of Lubrication and Tribology, Volume II*, p. 1169, CRC Press, Boca Raton, FL, USA, 2012.
146. Mobarak, H.M., Mohamad, E.N., Masjuki, H.H., Kalam, M.A., Al Mahmud, K.A.H., Habibullah, M., Ashraful, A.M., The prospects of biolubricants as alternatives in automotive applications. *Renew. Sustain. Energy Rev.*, 33, 34–43, 2014.
147. Rudnick, L.R., *Synthetics, Mineral Oils, and Bio-Based Lubricants*, 2nd ed, CRC Press, Boca Raton, FL, USA, 2013.
148. Papay, A.G., Antiwear and Extreme-Pressure Additives in Lubricants. *Lubr. Sci.*, 10, 209–224, 1998.
149. American Petroleum Institute (Ed.), *Appendix E—API Base Oil Interchangeability Guidelines for Passenger Car Motoroils and Diesel Engine Oils*, pp. 1–34, American Petroleum Institute, Washington, DC, USA, 2019.

150. Speight, J.G., Chapter 3: Hydrocarbons from crude oil, in: *Handbook of Industrial Hydrocarbon Processes*, pp. 95–142, Elsevier, Amsterdam, The Netherlands, 2020.
151. Stachowiak, G.W. and Batchelor, A.W., Chapter 3: Lubricants and their composition, in: *Engineering Tribology*, 3rd ed, Elsevier, Amsterdam, The Netherlands, 2006.
152. Wu, M.M., Ho, S.C., Forbus, T.R., Synthetic lubricant base stock processes and products, in: *Practical Advances in Petroleum Processing*, Springer, New York, NY, USA, 2006.
153. Nowak, P., Kucharska, K., Kaminski, M., Ecological and Health Effects of Lubricant Oils Emitted into the Environment. *Int. J. Environ. Res. Public Health*, 16, 3002, 2019.
154. Salimon, J., Salih, N., Yousif, E., Biolubricants: Raw materials, chemical modifications and environmental benefits. *Eur. J. Lipid Sci. Technol.*, 112, 519–530, 2010.
155. Reeves, C.J., Siddaiah, A., Menezes, P.L., A review on the science and technology of natural and synthetic biolubricants. *J. Bio-Tribo-Corros.*, 3, 1–27, 2017.
156. Dai, W., Kheireddin, B., Gao, H., Liang, H., Roles of nanoparticles in oil lubrication. *Tribol. Int.*, 102, 88–98, 2016.
157. Yang, G., Zhang, Z., Zhang, S., Yu, L., Zhang, P., Hou, Y., Preparation and characterization of copper nanoparticles surface-capped by alkanethiols. *Surf. Interface Anal.*, 45, 1695–1701, 2013.
158. Padgurskas, J., Rukuiza, R., Prosycevas, I., Kreivaitis, R., Tribological properties of lubricant additives of Fe, Cu and Co nanoparticles. *Tribol. Int.*, 60, 224–232, 2013.
159. Asadauskas, S.J., Kreivaitis, R., Bikulcius, G., Griguceviciene, A., Padgurskas, J., Tribological effects of Cu, Fe and Zn nano-particles, suspended in mineral and bio-based oils. *Lubr. Sci.*, 28, 157–176, 2016.
160. Flores-Castañeda, M., Camps, E., Camacho-López, M., Muhl, S., García, E., Figueroa, M., Bismuth nanoparticles synthesized by laser ablation in lubricant oils for tribological tests. *J. Alloys Compd.*, 643, S67–S70, 2015.
161. Chou, R., Battez, A.H., Cabello, J.J., Viesca, J.L., Osorio, A., Sagastume, A., Tribological behavior of polyalphaolefin with the addition of nickel nanoparticles. *Tribol. Int.*, 43, 2327–2332, 2010.
162. Peng, D.X., Yuan, K., Chen, S.-K., Shu, F.-C., Chang, Y.P., Dispersion and tribological properties of liquid paraffin with added aluminum nanoparticles. *Ind. Lubr. Tribol.*, 62, 341–348, 2010.
163. Alves, S.M., Barros, B.S., Trajano, M.F., Ribeiro, K.S.B., Moura, E., Tribological behavior of vegetable oil-based lubricants with nanoparticles of oxides in boundary lubrication conditions. *Tribol. Int.*, 65, 28–36, 2013.
164. Wu, H., Zhao, J., Cheng, X., Xia, W., He, A., Yun, J.-H., Huang, S., Wang, L., Huang, H., Jiao, S. *et al.*, Friction and wear characteristics of TiO_2

nano-additive water-based lubricant on ferritic stainless steel. *Tribol. Int.*, 117, 24–38, 2018.
165. Wu, H., Zhao, J., Xia, W., Cheng, X., He, A., Yun, J.-H., Wang, L., Huang, H., Jiao, S., Huang, L. *et al.*, A study of the tribological behaviour of TiO_2 nano-additive water-based lubricants. *Tribol. Int.*, 109, 398–408, 2017.
166. Laad, M., Ponnamma, D., Sadasivuni, K.K., Tribological studies of nano-modified mineral based multi-grade engine oil. *Int. J. Appl. Eng. Res.*, 12, 2855–2861, 2017.
167. Luo, T., Wei, X., Huang, X., Huang, L., Yang, F., Tribological properties of Al2O3 nanoparticles as lubricating oil additives. *Ceram. Int.*, 40, 7143–7149, 2014.
169. Peña-Parás, L., Taha-Tijerina, J., Garza, L., Maldonado-Cortés, D., Michalczewski, R., Lapray, C., Effect of CuO and Al_2O_3 nanoparticle additives on the tribological behavior of fully formulated oils. *Wear*, 332-333, 1256–1261, 2015.
170. Rabaso, P., Ville, F., Dassenoy, F., Diaby, M., Afanasiev, P., Cavoret, J., Vacher, B., Le Mogne, T., Boundary lubrication: Influence of the size and structure of inorganic fullerene-like MoS2 nanoparticles on friction and wear reduction. *Wear*, 320, 161–178, 2014.
171. Xu, Y., Hu, E.-Z., Hu, K.-H., Xu, Y., Hu, X., Formation of an adsorption film of MoS2 nanoparticles and dioctyl sebacate on a steel surface for alleviating friction and wear. *Tribol. Int.*, 92, 172–183, 2015.
172. Rosentsveig, R., Gorodnev, A., Feuerstein, N., Friedman, H., Zak, A., Fleischer, N., Tannous, J., Dassenoy, F., Tenne, R., Fullerene-like MoS2 nanoparticles and their tribological behavior. *Tribol. Lett.*, 36, 175–182, 2009.
173. Gulzar, M., Masjuki, H., Varman, M., Kalam, M., Mufti, R., Zulkifli, N., Yunus, R., Zahid, R., Improving the AW/EP ability of chemically modified palm oil by adding CuO and MoS2 nanoparticles. *Tribol. Int.*, 88, 271–279, 2015.
174. Zhou, L.H., Wei, X.C., Ma, Z.J., Mei, B., Anti-friction performance of Fe nanoparticle synthesized by biological method. *Appl. Surf. Sci.*, 407, 21–28, 2017.
175. Peng, D., Kang, Y., Hwang, R., Shyr, S., Chang, Y., Tribological properties of diamond and SiO2 nanoparticles added in paraffin. *Tribol. Int.*, 42, 911–917, 2009.
176. Raina, A. and Anand, A., Lubrication performance of synthetic oil mixed with diamond nanoparticles: E_ect of concentration. *Mater. Today Proc.*, 5, 20588–20594, 2018.

177. Gupta, M.K. and Bijwe, J., A complex interdependence of dispersant in nano-suspensions with varying amount of graphite particles on its stability and tribological performance. *Tribol. Int.*, 142, 105968, 2020.
178. Sivakumar, B., Ranjan, N., Sundara, R., Kamaraj, M., Tribological properties of graphite oxide derivative as nano-additive: Synthesized from the waster carbon source. *Tribol. Int.*, 142, 105990, 2020.
179. Phiri, J., Gane, P., Maloney, T.C., General overview of graphene: Production, properties and application in polymer composites. *Mater. Sci. Eng. B*, 215, 9–28, 2017.
180. Eswaraiah, V., Sankaranarayanan, V., Ramaprabhu, S., Graphene-based engine oil nanofluids for tribological applications. *ACS Appl. Mater. Interfaces*, 3, 4221–4227, 2011.
181. Lin, J., Wang, L., Chen, G., Modification of graphene platelets and their tribological properties as a lubricant additive. *Tribol. Lett.*, 41, 209–215, 2010.
182. Zhao, J., Mao, J., Li, Y., He, Y., Luo, J., Friction-induced nano-structural evolution of graphene as a lubrication additive. *Appl. Surf. Sci.*, 434, 21–27, 2018.
183. Wang, X., Zhang, Y., Yin, Z., Su, Y., Zhang, Y., Cao, J., Experimental research on tribological properties of liquid phase exfoliated graphene as an additive in SAE 10W-30 lubricating oil. *Tribol. Int.*, 135, 29–37, 2019.
184. Vats, B.N. and Singh, M., Evaluation of tribological properties of graphene oxide dispersed paraffin oil. *Mater. Today Proc.*, 25, 557–562, 2020.
185. Zhang, X., Zhu, S., Shi, T., Ding, H., Bai, Y., Di, P., Luo, Y., Preparation, mechanical and tribological properties of WC-Al2O3 composite doped with graphene platelets. *Ceram. Int.*, 46, 10457–10468, 2020.
186. Gulzar, M., Masjuki, H.H., Kalam, M.A., Varman, M., Zulkifli NW, M., Mufti, R.A., Zahid, R., Yunus, R., Dispersion stability and tribological characteristics of TiO2/SiO2 nanocomposite-enriched biobased lubricant. *Tribol. Trans.*, 60, 670–680, 2016.
187. An, V.V., Anisimov, E., Druzyanova, V., Burtsev, N., Shulepov, I.A., Khaskelberg, M., Study of tribological behavior of Cu-MoS2 and Ag-MoS2 nanocomposite lubricants. *SpringerPlus*, 5, 72, 2016.
188. Wang, Z., Ren, R., Song, H., Jia, X., Improved tribological properties of the synthesized copper/carbon nanotube nanocomposites for rapeseed oil-based additives. *Appl. Surf. Sci.*, 428, 630–639, 2018.
189. Shah, R., Das, M., Diloyan, G., Innovative use of nano particles to develop an Environment-Friendly Water-based High-Performance Lubricant. *Petro-chemical Energy*, 2023. https://www.petro-online.com/article/analytical-instrumentation/11/koehler-instrument-company/innovative-use-of- nano-particles-to-develop-an-environment-friendly-water-based-high- performance-lubricant/3251.
190. Saini, V., Bijwe, J., Seth, S., Ramakumar, S.S.V., Augmenting the lubrication performance of nano-oils through synergistic co-functioning of nanoparticles. *Tribol. Int.*, 182, 108332, 2023.

3

Characterization Techniques for Nanolubricants Using Different Approaches

Priyanka Chhabra[1]* and Akshara Johari[2]

[1]Centre for Medical Biotechnology, Amity Institute of Biotechnology, Amity University, Greater Noida, Uttar Pradesh, India
[2]Division of Forensic Science, School of Basic and Applied Sciences, Galgotias University, Greater Noida, Uttar Pradesh, India

Abstract
The extensive application of nanotechnology concerned with the synthesis and characterization of nanomaterials, nanoparticles, and nanostructures. These nanostructures when compared with bulk material exhibit different physical and chemical properties. These nanostructures repeatedly require sophisticated techniques for characterization. The characterization of nanomaterials is a fundamental pillar for the development and introduction of nanomaterials for specific applications. Therefore, the characterization of nanomaterials and nanostructures has been largely based on certain critical advancement of conventional characterization methods like X-ray diffraction (XRD), which has been widely used for the determination of crystalline character, crystallite size, crystal structures, and lattice constants of nanoparticles, nanowires, and thin films. Scanning electron microscopy (SEM) and transmission electron microscopy (TEM), together with electron diffraction, have been commonly used in the characterization of nanoparticles to get an idea of the size, shape, and defects present in these materials. In this chapter, we will describe briefly the tools and techniques that are widely used for the characterization of nanomaterials.

Keywords: Nanolubricants, nanoparticles, nanoadditives, synthesis, crystal structures

**Corresponding author*: pchhabra188@gmail.com

Mohd Yusuf, Lalit Prasad and Shafat Ahmad Khan (eds.) Nanolubricants: Generation and Applications, (61–88) © 2024 Scrivener Publishing LLC

3.1 Introduction

Lubrication is the practice or method of applying lubricant to minimize friction and wear between two moving surfaces. Lubrication has several advantages, including the avoidance of corrosion, water, and dust, as well as acting as an insulator in transformers. Wear and friction can lead to machinery failure and energy losses (for example, in the engine, shaft, bearings, and gears) [1]. As a result, sufficient lubrication is required to solve these problems. According to a study, friction in an engine can waste up to one-third of the fuel energy utilized [2]. Lubricant oil is a possible replacement for mineral oil, as it is biodegradable and harmless and does not pollute the environment in the same way as mineral oil. Pollution of aquatic and terrestrial ecosystems produced by mineral oil discharge had a direct impact on the environment [2, 3]. Nanoparticles are being developed as a new class of lubricant additives to address the difficulty of adding them to existing lubricants. Because it does not require tribo-active ingredients such as chlorine, phosphorus, or sulfur, which are hazardous to the environment, most NPs are ecologically benign and increase tribological qualities in lubricants, meaning they are more environmentally friendly [3].

Many studies have demonstrated the usage of various types of NPs as lubricant additives. Wear and friction reduction are great in metals, metal sulfides, metal oxides, boron nitrides, carbon materials, nanocomposites, and rare earth compounds. Lubricants containing nanoparticles as additives can offer improved lubrication efficiency and anti-wear and friction-reduction qualities, making them ideal for use in cars and motorbikes [4]. Nanoparticles can also impact friction forces and lubrication by changing the slip conditions at the fluid–solid interface. Metal nanoparticles have been investigated as severe pressure and anti-wear additives in lubricating systems. High pressures and temperatures can produce friction and surface damage, which they can alleviate. The size, shape, and concentration of nanoparticles, as well as their actuation method, all affect their friction reduction and anti-wear performance [5]. The addition of nanoparticles improves the tribological property of the lubricants. The study related to friction, wear, and lubrication is called tribology. Characterization of physical and chemical properties of the nanomaterials can be performed using various techniques and tools. Different factors affect the reactivity of the nanolubricants like rheological characters, kinematic viscosity, and volume fraction. The statistical analysis helps to determine the particle size. Morphology, stability validation, viscosity index, and kinematic viscosity are the important parameters to characterize the physical and chemical properties of nanolubricants [6]. To evaluate the influence of nanoparticles and the effect of their properties on addition to base oil is the main purpose of characterization. Characterization is mainly focused on nanotribology and rheology of the nanoparticles. This chapter includes a description of the various tools and techniques that are applied for the characterization of the nanolubricants and nanoparticles used as additives in the lubricants [7]. Factors affecting the characterization of nanomaterials are also briefly described in this chapter. Since nanotechnology is an emerging platform, it has its influence in every field of science and experiments.

3.2 Nanoparticles as an Additive to Nanolubricants

Improving lubricant efficiency will have a big influence on efforts to make manufacturing more sustainable and save energy. Efficient lubricants, for example, extend the life of industrial machinery, resulting in more efficient material consumption. The development of new lubricants raises frictional and wear-reduction concerns, as well as environmental concerns [8]. For example, today's petroleum-based lubricants were created at a time when wear reduction was considered more important than environmental concerns. Today's additives have a high potential for bioaccumulation and environmental toxicity, making waste oil a severe environmental problem combined with metal oxide powder after usage of high pressures and temperatures can produce friction and surface damage, which they can alleviate. Metal nanoparticles have recently been investigated as severe pressure and anti-wear additives in lubricating systems [9]. According to some studies, incorporating natural nanoparticles (NNPs) into lubricants improves tribological properties while also lowering the environmental impact of the lubricant. Nanoparticles are being used as lubricant additives in the oil and gas industry due to their unique qualities, such as high temperature stability, chemical inertness, high strength, and wear performance [10, 11]. Due to this, they have sparked a lot of interest in the advanced engineering area. Some of the most prominent classes of nanoparticles are carbon-based nanoparticles, metallic alloy nanoparticles, and ceramic nanoparticles. The most common nanoparticles, which are utilized as a dispersoid in nanolubricants, are molybdenum disulfide (MoS_2), hexagonalboron nitride (h-BN), and titanium dioxide (TiO_2). Reduction of rolling force and enhancement of the surface layer of the stainless steel was improved by developing a new type of lubricant, which is based on oil–water (O/W) mixed with nanolubricant containing the TiO_2 nanoparticles [12]. Addition of the nanoparticles affects the tribological behavior of the steel.

Nanoparticles can also be mixed as a lubricant condition. It symbolizes the wear, and the coefficient of friction decreases with the decrease in particle size. ZnO nanoparticles, when added, reduces wear and extreme pressure properties. Addition of Al_2O_3 nanopowder results in the enhancement and change of the properties like hardness and surface roughness [10, 13]. The tribological characteristics of the aluminum alloy covered with SiN and TiN film were improved by the addition of different types of nanoparticles to the lubricant oils. It also resulted in improvement of friction and wear rate. Lubricants are made from three types of basic oils: mineral, synthetic, and biological—all of which have various properties and are suited for different uses. Mineral oil is the most widely used lubricant in the business; it is the type of oil that is used to make petroleum-based fluids, which are used in machines [14]. Biological lubricants are commonly used in the food and pharmaceutical industries, such as bread ovens and kilns, where contamination is a concern. Vegetable oils include palm, rapeseed, and castor, whereas animal oils include sperm, fish oil, and lanolin. Selection of base oil or base fluid decides the performance of the lubricant and its properties. The oils vary from mineral oils to synthetic oils. Mineral oils are made by fractional distillation and refining of crude oil, and are the most often used basic stock [15]. Removal of the high molecular weight of paraffin is done to minimize wax deposition while lubricating and aromatic hydrocarbons to slow down the reduction in oil viscosity.

The advantages of mineral oils are availability and low cost, but solidification at low temperature, viscous loss, and oxidation are some of the major disadvantages. The use of synthetic lubricants has progressively risen over the previous century, particularly for applications that could not tolerate mineral oils [16]. There are several different hydrocarbon fluids that may be used as lubricants. Production economics, on the other hand, strictly limits their separate applicability ranges. Oils now synthesized by oil refineries are low-cost to create, making them cost-effective for uses such as engine oils. Silicones and silahydrocarbons are two important types of chemicals, both of which include silicon. Although both kinds of chemicals include silicon, their molecular structures are substantially different [17]. This category includes a variety of synthetic lubricants that have been shown to be chemically stable. Organohalogens are a kind of halogen that can be found in nature and have antioxidant properties. Machines that use pure sulfuric acid, such as air or oxygen compressors, were previously reliant on it. Fluorine and chlorine are employed to create compounds with good oxidation and thermal stability [18]. Perfluoropolyethers, chlorofluorocarbons, chlorotrifluoroethylenes are the most prevalent synthetic lubricants. Oxidation stability, anti-friction and wear and anti-corrosion, and stability against biological degradation can be improved by incorporating suitable additives [19]. The base fluid should have the ability to retain the additive in the solutions under all operating situations as a transporting agent for additives. Additive packages form almost 10% of the lubricant products. The number of studies on lubricants incorporating nanoparticles has expanded dramatically in recent years in order to decrease systemic wear and friction. Both organic and inorganic nanoparticles have been studied extensively for use as extreme pressure (EP) and anti-wear agents [20]. Types of the additives used are metal and oxides of metals. Researchers have been interested in the tribological characteristics of metal nanoparticles. Cu nanoparticles, for example, displayed great self-repairing characteristics in addition to tribological efficacy, while also being ecologically beneficial. Nanoparticles, which include Fe, Cu, and Co, as well as their combinations, were put to the test. To establish the wear impact of the aforementioned particles, certain surface experiments were carried out. Different experiments were used for the analysis of the influence of nanoparticles of Cu on the behavior of diesel oil with serpentine powder as its nanoparticle [21].

Sn and Fe nanoparticles were introduced to multialkylated cyclopentanes (MACs) to see what effect they had. In MACs oil, these nanoparticles decreased friction, wear, and heat. As a result of Fe's higher solubility in steel surfaces, its nanoparticles have a greater anti-wear capacity. Metal oxides are often employed as additives in lubricant base fluids, with the resultant mixture being utilized for anti-friction and anti-wear purposes. Nano-$TiO_{2,}$ nano-ZnO, nano-$ZnAl_2O_4$ are the most commonly used nanomaterials to be used as lubricant additives for enhancement. With the application of TiO_2 microcrystals, the lubricant's load-bearing capacity, anti-wear, and friction reduction initially improved, but eventually decreased for larger concentrations [21]. The concentration and kind of nanoparticles affect the rheological behavior of the nanolubricant. The viscosity of oils improved. SiO_2 nanoparticles rose as the concentration of SiO_2 nanoparticles grew, whereas the viscosity of oils enhanced with TiO_2 nanoparticles dropped as the concentration of TiO_2 nanoparticles increased [22, 23]. When SiO_2 and TiO_2 were added to oil, different rheological behaviors were found. Shear thickening was seen in oils boosted with SiO_2 nanoparticles, whereas shear thickening was observed in sunflower oil increased with TiO_2 nanoparticles. Among those additives, nano-TiO_2 demonstrating as one of the best candidate materials has drawn significant attention due to its low cost, nontoxicity, superb dispersibility, and stability in base lubricant, excellent lubricating properties; and practical potentiality in engineering applications [23]. Nanoadditives are a type of additive used in a variety of products to reduce friction and wear on sliding surfaces. Nanolubricants have been extensively researched as an alternative to traditional lubricating oils, since they allow for significant friction reduction and unique properties. Load-carrying capacity has improved.

3.3 Application of Nanolubricants

As biolubricant oils have certain technical characteristics that are derived from nature and are biodegradable, biolubricant oils are viewed as alternatives to mineral oils. The rise in oil costs, the depletion of the world's crude oil reserves, and the need to safeguard the environment are all factors that have rekindled interest in their development and use [24, 25]. Due to the long fatty acid chains and polar groups included in the structure of vegetable oil, biolubricants can offer both boundary and hydrodynamic lubrication. Vegetable oil is made from oil-bearing seeds that are widely available across the world [26]. There are 350 oil-bearing crops accessible globally, according to studies. Vegetable oil can be eaten or it cannot be eaten. Jatropha, karanja, neem, rice bran, castor, linseed, mahua, palm, sunflower, coconut, soybean, olive, and canola are all examples of plants that can be used to make oil. Many researchers have said that vegetable oil can be used as engine fuel, but a few have also said that it can be used as car oil [27].

As a different kind of lubricant for industrial and maintenance uses, biolubricants offer a lot of benefits. Because they are sensitive, they help protect the environment and keep it from getting dirty. Biolubricants can be employed in a variety of situations in industry and maintenance. Due to their low toxicity, good lubricating properties, high viscosity index (VI), high ignition temperature, longer equipment service life, and high load-carrying capacity, biolubricants could be used as an alternative to oil in cars [28]. They also offer a number of other advantages over traditional lubricants, such as high resistance to corrosion, good anti-wear properties, and a high coefficient of friction, and low evaporation rates, friction, and natural multi-grade qualities. Due to its better intrinsic properties, oil has the potential to be developed commercially because it offers considerable advantages as an alternative lubricant. It is suitable for a variety of industrial and maintenance applications [29, 30]. Some of the most important things they are used for are metalworking fluids, machine oils, compressor oils, hydraulic oils, and special oils. Even though it appears that all system elements have already attained their maximum efficiency, performance enhancements in heating, ventilation, and air conditioning (HVAC) applications are a strict necessity in the business. In this regard, nanotechnologies, particularly nanofluids, might provide intriguing possibilities. Calcium carbonate nanoparticles were used as additions in a poly-alpha-olefin to boost load-carrying capacity while also providing antiwear and friction-reducing qualities [30]. The lubricating performance of water-based nanolubricants including nano-TiO_2 in hot steel rolling was investigated in terms of rolling force, surface roughness, and rolled steel oxide scale thickness, with a suggested lubrication mechanism. There are many other several applications of the nanolubricants in engineering and technical engineering science.

3.4 Preparation of Nanolubricants

In order to enhance the wear and friction properties of the lubricant's different nanoadditives, these additives are synthesized using different methods. Nanoparticles with a diameter of 40 nm (Al_2O_3) and densities of 2.2 g/cm^3 (SiO_2) were acquired from the market by researchers at the European Society for the Electromechanical and Electrochemistry (ESEM) in Paris [31]. The diameter of nanoparticles is 40–46 nm, which corresponds to the supplier's stated size (40 nm). Although the nanoparticles are spherical, SiO_2 nanoparticles have crystalline and amorphous structures, and so forth. Because the moisture content of nanoparticles influences their lubricant miscibility, they are dried in an oven at 80°C to 100°C for half an hour before being weighed and dispersed in the base oil. Plain gear oil and gear oil nanolubricants were dispersed individually with Al_2O_3 and SiO_2 nanoparticles at concentrations of 0.3%, 0.6%, and 0.9% nanoparticle volume fractions in a batch of seven samples. Graphite is combined with twice the weight of nanoparticles and oleic acid to make a nanolubricant [31]. The mixture is then placed in an ultrasonic wash for roughly 3 h, with a 15-min break every 30 min, and the technique of preparation. O/W, 0.5 TiO_2 nanoparticles with a sphere-like particle size of 30 nm were combined with a lubricating oil with a viscosity of 1.20 poise and a density of 0.90 g/cm^3 at 23°C to create nanolubricants. During the hot-rolling process, they can function as a transporter of nanoparticles and a source of oil film in the contact region.

CuO nanoparticles were introduced to base oils at 0.1 wt% concentrations during the nanolubricant production. The weight of the nanoparticles is determined by the amount of base oil used and the proportion of nanoparticles utilized. An electromagnetic stirrer was utilized for mixing the nanoparticle with the surfactant and base oil. Lubricant additives are compounds that are added to the basic oil to improve wear and friction properties by allowing for adsorption. N-rGO was used to create nanolubricants, which were determined by probing the settling after 1 month and analyzing their stability. N-rDO was created by using a probe sonicator for 30 min at a frequency of 20 kHz and a power of 500 W (20%) to create different levels of nanoparticles in oil samples [32]. SiO_2 nanoparticles were used to make a nanolubricant consisting of SiO_2 particles dispersed in polyalkylene glycol (PAG) lubricant and an ultrasonic bath vibrator. The optimal duration for sonification was found using UV-visible spectrophotometer measurement to ensure that the nanoparticles are thoroughly mixed with decreased sedimentation potential and a better mean particle size. The ideal duration for sonification—the process that ensures nanoparticles are thoroughly mixed with decreased sedimentation potential and a better mean particle size—has been determined by researchers at the University of Bristol in the UK [33, 34]. The sonicator will break down the agglomeration and ensure that the nanoparticles properly disseminate in the base solution. The nanolubricants were made using a two-step approach of dispersing carbon/graphite (CG) nanoparticles in palm kernel oil (PKO). CG nanoparticles with a concentration of 0.05 wt% are dispersed in PKO using three distinct two-step processes to investigate the impacts of the nanolubricant's dispersion method (magnetic stirrer, overhead stirrer, and high-shear homogenizer). SiO_2 nanoparticles were scattered by an ultrasonic probe sonicator with a fixed frequency of 20 kHz, followed by a prolonged bath sonication using a water bath. Urea hydrolysis and deionized water were used to make layered double-hydroxide (LDH). The green powders were produced by vacuum drying for 12 h at 100°C. They were then combined with $Ni(NO_3)_2 6H_2O$ (8.6 g), $Al(NO_3)_3 9H_2O$ (4.3 g), and urea (30 g) for 24 h at 92°C [34]. After that, the material was filtered and rinsed with deionized water. A hydrothermal reaction was carried out at 99 W for 140 s with ON/OFF pulses of 20 and 10 s, which were repeated nine times to keep the solution at 80°C.

CuO nanoparticles were made utilizing a microwave process with a copper acetate alcoholic solution and a sodium hydroxide solution. After this, the solutions were chilled for 5 min at 15°C. Centrifugation at 3600 rpm for 5 min was used to separate the nanoparticles from the suspensions. The nanoparticles were cleaned in hot water and ethanol many times. They were then dried at 60°C for 2 h or until all the ethanol had evaporated. The synthesis was carried out three times. According to literature surveys, apart from these described methods, there are other several methods which used synthesis of nanolubricants. Nanolubricants are mainly prepared dispersing synthesized nanoparticles in common non-polar solvents and mineral oils to provide a stable and transparent solution [15]. Most nanolubricants have been made from synthetic-based lubricants and have been characterized for their anti-wear and anti-friction properties (rheological behavior of zinc-oxide nanolubricants).

3.5 Characterization Factors of Nanolubricants

Nanolubricants have a number of properties that affect how well they transfer heat from one place to another, such as their viscosity, density, thermal conductivity, and specific heat capacity. The fluctuation of thermophysical characteristics with temperature is discussed. Nanolubricants have unique tribological characteristics. The majority of the investigations in the literature show that temperature and nanoparticle concentration are critical elements in affecting the viscosity of nanofluids. Nanolubricants have sparked a lot of interest due to their unique features and the potential influence they could have on the energy and automobile industries. Understanding the influence of nanolubricants on shear rate and shear stress has been aided by the rheological behavior of nanolubricants. The viscosity of nanolubricants plays an important role in determining their tribological qualities. The addition of nanoparticles to engine oil has been found to minimize friction and wear of engine parts that come into contact during movement, as well as the mechanism. Rheology is the use of a rheometer to examine the flow and deformation of materials under applied forces. Rheological qualities can be measured with a rheometer from the bulk to the nanoscale. Rheological qualities can be determined at the bulk scale using a mechanical rheometer, and at the nanoscale with a capillary viscometer. The viscosity of five distinct materials was determined in this experiment using an Anton Paar Rheolab QC Rheometer and Rheolab software. The Rheolab QC is a rotational rheometer with a Peltier temperature device that may be used to measure temperature (0–180°C). Tribological performance is dependent on lubrication, friction, and wear. The creation of a protective layer or tribofilm on contact surfaces is what lubricity is all about. High lubricity lowers friction and energy losses by eliminating direct surface contact. Having said that, because the creation of a lubricant film does not necessarily imply superior wear protection, a high lubricity is not always accompanied by higher wear protection. The absorption of the surface-active material results in a protective layer on the rubbing surfaces. Basic stocks or additives are examples of lubricants. There are three forms of surface asperity that come into contact with each other. As a result of mechanical wear, the following may occur: adhesion, abrasion, and fatigue, and these are all terms used to describe how something sticks to something. Other than the above characteristics, particle size also matters. Nanoparticles have a characteristic size of less than 100 nm, such as the diameter of a spherical or tubular particle or the thickness of a lamellar. They may easily access the contact surface for load bearing and lubrication due to their tiny size. A high surface-to-volume ratio is also beneficial. The frictional behavior of nanoparticles is determined by the interaction between neighboring layers. Tribological performance is said to be influenced by the number of layers and interlayer spacing. When the number of layers is reduced, other concerns arise, such as puckering and wrinkles, inclination angles, and interlayer spacing.

3.6 Characterization Techniques Used for Nanolubricants

There are different techniques used for the characterization of nanolubricants. Techniques used for characterization can be differentiated based on different factors. The factors affecting characterization includes morphology (shape/size), topography (surface), chemical properties, electrical properties, and optical properties.

3.6.1 Morphology and Topography Analysis

The most basic set of information is the shape and size of the particle. The suitability of the nanoparticles is determined for the specified applications.

3.6.1.1 Dynamic Light Scattering (DLS): Particle Size Analysis

The DLS method is based on how Rayleigh scattering changes the intensity of the light source, and the results are often given as changes in intensity as a function of particle diameter. Dynamic light scattering is the method most often used to measure the size of particles in a colloidal environment (photon correlation spectroscopy or quasi-elastic light scattering). The DLS method is useful for determining the diameter of nanoparticles that are scattered in a fluid media. Particles of various chemical natures, such as polymeric, metallic, and oxide, can be investigated. Furthermore, this technique may be used to determine the size of coated particles. Brownian motion, or the outcome of a thermically produced collision between the particles and the solvent molecules, is the basis for measuring light scattered by particles to quantify their size. When a laser source bombards colloidal particles distributed in liquid media, variations in the intensity of the light scattered as a function of time are observed. The profile of these changes is influenced by particle size, or, to put it another way, by the diffusion coefficient. Larger particles travel at a slower rate, resulting in greater time scale variations in the intensity profile of light scattering. The DLS size corresponds to the diameter of a sphere traveling at the same coefficient of diffusion (D) as the particles under investigation. The surface structure and concentration of the particles, as well as dispersion medium factors such as ionic strength, viscosity, and refractive index, can alter the value of diffusion coefficient (D), enabling the investigation of particle surface characteristics. The size examination of particles in the submicrometer range is the most common application of DLS. It takes use of the Brownian motion of the particles, which produces a permanent spatial rearrangement and hence, influences the interference of the individual scattering signals, resulting in a fluctuating signal at the detector. The particle diffusion coefficient Dp, which is inversely proportional to particle size, determines the kinetics of the rearrangement.

$$D_p = \frac{k_b T}{3\pi\eta x_{h.t}}$$

The Stokes–Einstein equation has k_b and T as Boltzmann's constant and temperature, respectively, while $x_{h,t}$ is the fluid's dynamic viscosity and $x_{h,t}$ is the hydrodynamic diameter of translational motion. The sample concentration range for optimal measurements in DLS is heavily influenced by the sample materials and their size. There may not be enough scattering events to make a reliable measurement if the sample is too dilute. Multiple scattering, on the other hand, can occur if the sample is overly concentrated.

Furthermore, at large concentrations, the particle may not be freely mobile, with its spatial displacement being influenced by particle interactions rather than Brownian motion. This is especially true in the case of magnetic nanoparticles (MNPs) with magnetic dipole–dipole interactions between the particles. The hydrodynamic size and colloidal stability of magnetic nanoparticles with spherical or anisotropic shapes are monitored using dynamic light scattering. DLS delivers statistically representative data on nanomaterial hydrodynamic size. The intensity of scattered light is monitored in DLS by particles in a colloid dispersion scattering an incident laser. Constantly moving dispersion particles cause constructive and destructive interferences, causing the dispersed light's intensity to fluctuate over time. In DLS, intensity fluctuations in scattered light are linked with short decay periods, and the intensity autocorrelation function (ACF) is calculated using the following monoexponential equation.

$$G(\tau)= 1+b.e^{-2Dtq2\tau}$$

DLS is the suitable instrument for the size measurement of nanoparticles. DLS findings are influenced by a number of factors, including the solvent's viscosity, the instrument's temperature, and the material's refractive index (RI). The amount of light scattered by distributed NPs is related to the sixth power of their radii. When the particles are $1/10^{th}$, the wavelength of the input light, the scattered light has the same energy as the incident light and is not angle-dependent (elastic scattering) (Rayleigh scattering). When the particle size exceeds /10, however, Rayleigh scattering is replaced by anisotropic Mie scattering, in which the scattered light has an energy difference from the incoming light (inelastic scattering) and is angle-dependent where the scattered light is most intense towards the incident light's direction Because of the way electromagnetic waves (e.g., light) interact with particles, this size threshold is outside the scope. While the particles in colloidal dispersion travel in a random and translational manner, they also spin at a high rate. Interparticle interactions are particularly crucial, as the frequency of collisions inside particles rises as the concentration increases, but the average path length covered by the particles between subsequent collisions decreases. Depending on the adsorbed layer, the surfaces of the dispersed particles are varied. The protein absorption on the surface of NPs, which results in the distinctive protein corona, is a frequent example of this phenomenon. The dispersed particles have a hydrated surface that is encased in a cloak of molecules that are not the particles' components. The corona is frequently discovered to be made up of both 'hard' and 'soft' components. The particles used in DLS have a different composition and surface chemistry from the ones that were initially manufactured. Modern DLS instruments contain Avalanche Photo Diode (APD) detectors with a quantum efficiency of 65% in red wavelengths, therefore, lasers of 633 nm are employed. The detectors in the most recent instruments are angled at 173 degrees to detect backscattering, however, the angle is still 90 degrees in certain older models (Nano S90, ZS90). Backscattering may be detected and excess scattered light can be avoided by positioning detectors at 173 degrees.

3.6.1.2 Electron Microscopy

For determining the features of the nanoparticle form and surface, electron microscopy (scanning EM/tunneling EM) can be used. The utilization of biological samples (nucleic acid coated particles) is limited by the technical and time-consuming sample preparation methods connected with them. Electron microscopy pictures of biomolecule-coated particles are of low quality. Dehydration causes sample shrinkage. As a result of the tendency of drying samples to agglomerate, more artefacts are added to the mix. In addition, there are greater analysis durations and the inability to do multiple analyses. Another set of difficulties is calculating dispersion.

3.6.1.3 X-Ray Diffraction

The nature of nanoparticles was validated using XRD. The single-phase monoclinic structure of the nanoparticles is shown by the XRD diffraction pattern. In the range of 2 theta 20–80, X-ray diffraction is seen. The XRD spectrum was used to investigate the nanoparticles' crystalline nature, crystal phase, and crystallite size. The nanoparticle XRD spectrum acquired standard X-ray diffraction spectrum is used for different nanoparticles. The size and form of the nanoparticles are determined by the intensity of the XRD peaks, whether the particles are crystalline or amorphous including their size. The shape of the nanoparticles can also be determined using XRD; it also can be used for the analysis of the various other characteristics of nanoparticles like surface area, dislocation density, and morphology index. Nanoparticle size and size distribution are estimated using X-ray diffraction (XRD). XRD is a commonly used technique for determining the crystallinity of synthesized nanoparticles. As a result, using XRD profile analysis, an attempt was made to estimate the crystallite size and crystallite size distribution of the particles using XRD data. Scherrer linked the breadth of XRD profiles to the size range of diffracting crystallites. The size as determined by XRD peak broadening refers to the material's smallest unfaulted areas or coherently scattering domains. In actuality, this is the area defined by flaws and grain boundaries and separated from the surroundings by a tiny misorientation, usually one or two degrees.

3.6.1.4 Atomic Force Microscopy

Atomic force microscopy (AFM) has become a significant researcher's tool in the physical and chemical materials, and medicinal sciences since its invention in 1985. AFM primarily uses a sharp tip on the end of a microscale cantilever of a probe to sense or modify environmental materials. Next, the surface properties of samples or pre-designed nanostructures are assessed. The force between an AFM probe and a sample may be directly "felt" using the probe. An AFM probe is very sensitive to interactions at the atomic level and is meant to find them. The AFM probe usually has a sharp tip that is less than 5 m tall and less than 10 nm wide at the top. This sharp tip is at the end of a 100–500 m long microscale cantilever. A piezo scanner is used to adjust the probe-sample location accurately in both the vertical and horizontal directions. When a voltage is supplied to the piezo scanner, it bends, expands, and contracts in a regulated and predictable manner. A professional program for regulating the probe and piezo scanner's operational settings, it may also be used to show and evaluate the findings. A feedback control system that consists of a laser diode, a position-sensitive photodetector, a controller, a feedback circuit, and other components used for AFM contact mode scanning. The feedback control system uses an analog/digital converter to receive and alter the precise probe-sample location, both vertically and laterally. It connects the piezo scanner with the professional software. Choosing the right AFM probe is important for any AFM experiment to work. When an AFM tip scans over the surface of a sample, it can cause two main types of artefacts. A sample height compression effect caused by the sample's elastic deformation and a tip widening effect caused by the tip and sample twisting together.

There are two ways to make these artefacts less harmful. The surface information may be analyzed using surface/tip deconvolution methods. The other option is to try different probes to find one that is suited for the experiment. To avoid contact between the cantilever and the sample, the tip height should be at least double the sample characterization height. AFM probes that are often used include silicon nitride probes, silicon probes, and carbon nanotube tips. The tips of AFM probes can be made to do different things by coating them with polymers or proteins. Also, spheres can be glued together to make sphere tips (colloid, glass, etc.). How well an AFM works is also affected by how sharp the tip is. This is shown by the radius of curvature of the tip (Rtip), the tip aspect ratio, and the cantilever spring constant (k). Different AFM modes and methods can be used to do imaging, measure mechanical properties, and change the surface. AFM imaging can work in three basic ways, depending on what kind of force is being measured and how it is being measured. Most of the time, these three ways are used to find out how tall someone is. Also, based on the three main AFM imaging modes, some additional imaging modes have been made. Some of these are deflection imaging, phase imaging, lateral force imaging, magnetic force gradient imaging, torsional resonance imaging, conductive AFM, electric field gradient distribution imaging, surface potential imaging, electrical carrier concentration imaging, force modulation imaging, surface thermal imaging, and scanning spreading resistance imaging.

In general, the imaging operation mode depends on the surface features that are of interest and how hard or sticky the sample is. Common surface mechanical measurements include force spectroscopy, force volume, nanoindentation, lateral bending test, two-point bending test, and three-point bending test. Dip-pen nanolithography, nanografting, nanoshaving, wear testing, electrochemical AFM nanolithography, thermal AFM nanolithography, nanomanipulation, and other ways to change the surface are common. AFM is useful for materials science, nanoscience, medical science, biological science, the semiconductor industry, and other fields.

3.6.1.4.1 Applications of AFM for Nanomaterials

(1) Height Imaging. The most widely used application of AFM in the scientific area is height imaging. There are three primary AFM imaging modes of operations. Contact mode, noncontact mode, and tapping mode:

- Contact mode of imaging is an operation that uses an AFM tip in contact with the sample surface. Due to the sensitivity of the cantilever deflection, the surface topology of the sample changes when it comes in contact with AFM tip while scanning. To get the height image, the angle of the cantilever is kept the same. The contact mode is best for hard samples and is also used for high-resolution imaging.
- The oscillation (amplitude <10) of cantilever is near its resonance frequency, i.e., typically 100–400 kHz. When the sample is under the AFM tip, the movement of the scanner is gentle and the distance from the tip of the AFM and sample surface is 1–10 nm, which results to the desired height image. In noncontact mode, the operations should be careful to avoid jump-to-contact with the water layer between the sample and the AFM tip. The mechanical changes of interaction of the tip–sample are tracked using high performing Z-servo feedback system and Z-scanner. Noncontact mode allows to prevent the tip contact with the sample, hence, the sample is neither destructed nor disturbed by the AFM tip particularly the soft sample, which is the major advantage.
- The oscillation amplitude of the cantilever is generally >20 nm, which is close to its resonance frequency, i.e., typically 100–400 kHz. The intermittent contact is created during the movement of the AFM tip across the sample surface and the amplitude of oscillation is damped. The height

image is obtained at the constant dampness of the cantilever oscillation amplitude.

Analysis of the morphology of the nanomaterial can be done using height imaging. The analysis of the surface roughness of the nanomaterials can be done by height imaging.

(2) Deflection Imaging. Contact mode is the mode that is most likely to perform deflection imaging. The height is constant if the feedback system is off in contact mode, and hence, the obtained image is deflected image. So, this mode is known as constant mode or deflected mode. In combination with height imaging, it is used for the analysis of the surface property of the nanomaterials. It has been used mostly for 3D nanostructures and their topography.

(3) Phase Imaging. It is derivative of tapping or noncontact mode. Piezoelectric driver is used to excite the cantilever into resonance oscillation, and the oscillation is changed into tip–sample interaction. Phase imaging is the process of making a map of the difference between the oscillations of the cantilever and the signals made by the piezoelectric driver. Phase shift can be caused by phase imaging. Surface properties including adhesion, friction, electrical properties, and magnetism can be analyzed. This can be used to obtain valuable information of sample surface. Surface composition studies of polymers by ion modification, compositional differences on the heterogeneous nanostructure surface, quantitative characterization of phase separation in polymer-blended films, and other applications of phase imaging have already been widely used in nanomaterials research.

(4) Lateral Force Imaging. Lateral force microscopy (LFM), also known as friction force microscopy, is used to map the relative differences in friction force between the tip and the sample surface. LFM may capture both the height and lateral force images at the same time. Contact mode AFM is comparable to LFM. In contact mode AFM, scanning is parallel to the long axis of the cantilever, and the force between the tip and the sample surface causes the cantilever to bend up and down. This is done to get information about the surface properties. While scanning is perpendicular to the cantilever's long axis in LFM, the force between the tip and the surface of the sample twists the cantilever around its long axis. This keeps the cantilever's sideways movement, which is used to measure the surface's properties. Changes in surface qualities (hydrophilicity, hardness, etc.) or changes in topography might cause the cantilever to twist. The reason for the cantilever twisting may be determined by comparing the trace LFM picture, the retrace LFM image, and the matching height image (trace or retrace image). LFM is currently commonly utilized to evaluate nanomaterial surface friction differences and subsequently to assess nanomaterial composition and structural differences. Self-assembled layers of cyclodextrin–thiocholesterol inclusion complexes on gold were studied using LFM. Height photos revealed no visible height differences, whereas LFM images revealed clear phase separation, demonstrating that LFM has a clear advantage when examining self-assembled layers made up of different components. At the nanoscale and microscale, LFM was utilized to distinguish between various phases of hydrated cement paste (calcium-silicate-hydrate particles, calcium hydroxide crystals, and unhydrated particles). As the temperature increased, the LFM signal dropped, indicating that LFM may be utilized to investigate the thermodynamic characteristics of nanomaterials.

(5) Magnetic Force Gradient Imaging. Magnetic force microscopy (MFM), which is developed from tapping mode, is used to image the magnetic force gradient. When a sharp magnetized tip is used to scan a magnetic sample, magnetic interactions between the tip and the sample are detected, allowing the magnetic force gradient to be mapped while the tip is above the sample surface (no touching). There are two forces acting on the tip and the sample surface during MFM measurements: magnetic and Van der Waals forces. As a result, MFM can measure both the height picture and the magnetic force gradient image concurrently using the "force range" or "two pass" approach. MFM has proven itself as a reliable workhorse in the fields of magnetic nanomaterials research and development. The magnetic response of these nanosheets was shown to be dependent on their layer number by examining phase and amplitude alterations. When MoS_2 and graphene nanosheets reached a particular thickness, they became nonmagnetic. The use of MFM brings up new possibilities for understanding the inherent characteristics of 2D nanomaterials.

(6) Torsional Resonance Microscopy. Torsional resonance microscopy, often known as torsional resonance microscopy (TRM), is a technique for measuring and controlling dynamic lateral forces between the tip and the sample. The cantilever oscillates in a twisting motion around its long axis in TRM, causing the tip to vibrate in a dithering motion. TRM can study the in-plane mechanical properties of nanomaterials like friction, shear stiffness, and other properties that are important to tribology. This is possible because both the vertical deflection and the lateral twist of the cantilever can be measured at the same time. TRM has better resolution, sensitivity, and stability than many other imaging modes. TRM demonstrated greater sensitivity and stability in water than traditional noncontact mode AFM, demonstrating that TRM can be useful for investigating soft matter in liquid and liquid–solid interfaces. Torsional resonance is also commonly used to image nanostructures in materials including human hair, mineral pennine, protein-DNA complexes, and polymers.

3.6.1.5 UV-Visible Spectroscopy

One of the most often used characterization techniques for determining particle production and characteristics is UV-visible spectroscopy. Furthermore, the size, shape, interparticle interactions, free electron density, and surrounding medium all impact the spectrum surface plasmon resonance of nanoparticles, indicating that it is an effective instrument for monitoring electron injection and aggregation of NPs. In contrast to a reference or blank sample, UV-visible spectroscopy is a type of analysis that looks at how many different wavelengths of UV or visible light a sample absorbs or lets through. This property is affected by the way the sample is put together, which could tell you what is in the sample and how much of it is there. Because this spectroscopic approach relies on the usage of light, we will start with the qualities of light. The quantity of energy contained in light is inversely proportional to its wavelength. As a result, shorter light wavelengths carry more energy while longer wavelengths carry less. To boost electrons in a substance to a higher energy state, which we can detect as absorption, a precise quantity of energy is required. In a substance, electrons in different bonding environments require a variable amount of energy to promote them to a higher energy state. This is why different wavelengths of light are absorbed by different substances. Humans can see a visible light spectrum ranging from 380 nm, which we perceive as violet, to 780 nm, which we perceive as red. UV light has wavelengths that are roughly 100 nm shorter than visible light. As a result, light may be defined by its wavelength, which can be useful in UV-visible spectroscopy for analyzing or identifying various compounds by locating the exact wavelengths that correlate to maximal absorbance Figure 3.1.

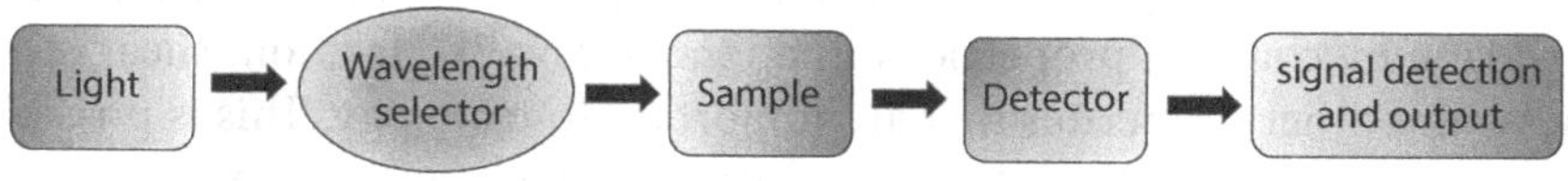

Figure 3.1 Schematic diagram of the main components in a UV-visible spectrophotometer.

The main components of the UV-visible spectroscopy:

- Light sources: A constant source capable of emitting light across a wide variety of wavelengths is required for this light-based approach. In both the UV and visible bands, a single xenon lamp is widely utilized as a high-intensity light source. Xenon lamps, on the other hand, have higher costs and are less stable than tungsten and halogen lamps. A tungsten or halogen lamp is usually used for visible light, whereas a deuterium lamp is commonly used for UV light in devices with two lamps. The light source of the device must swap during measurement since two distinct light sources are required to scan both the UV and visible wavelengths. In practice, this switchover usually happens between 300 and 350 nm, when the light emission from both light sources is similar and the transition can be smoother.
- Wavelength selection: Monochromators are devices that divide light into a small range of wavelengths. It is most commonly based on diffraction gratings that may be adjusted to select the appropriate wavelength of light by changing the incoming and reflected angles. The groove frequency of a diffraction grating is commonly expressed as the number of grooves per millimeter. A greater groove frequency improves optical resolution but limits the wavelength range that can be used. A lower groove frequency allows for a wider wavelength range to be used, albeit at the expense of optical resolution. UV-visible spectroscopy can use grooves with a density of 300 to 2,000 per millimeter, but a minimum of 1,200 grooves/mm is typical. Physical defects in the diffraction grating and the optical setup affect the quality of the spectroscopic observations. As a result, ruled holographic diffraction gratings have a higher defect rate than blazed holographic diffraction gratings. Holographic diffraction gratings with a blazed pattern tend to produce higher-quality observations. Absorption filters are often composed of tinted glass or plastic and are designed to absorb specific wavelengths of light. Interference filters, also known as dichroic filters, are made up of numerous layers of dielectric material with interference occurring between the thin layers of materials. These filters can be used as a wavelength selector by removing unwanted wavelengths through destructive interference. Cutoff filters allow light to

pass through either below (shortpass) or above (longpass) a given wavelength. Interference filters are routinely used to achieve this. Band pass filters are made up of shortpass and longpass filters that allow a wide range of wavelengths to pass through. Because of its versatility, monochromators are the most typically employed for this operation. Filters, on the other hand, are frequently used in conjunction with monochromators to further reduce the wavelengths of light selected for more exact measurements and improved signal-to-noise ratio.

- Sample analysis: The light then passes through a sample, regardless of whatever wavelength selector is utilized in the spectrophotometer. Measurement of a reference sample, sometimes known as a "blank sample," such as a cuvette filled with the same solvent used to create the sample, is required for all analyses. When an aqueous buffered solution containing the sample is used for measurements, the reference is an aqueous buffered solution that does not include the substance of interest. The sterile culture media would be used as a reference while assessing bacterial cultures. The equipment then uses the reference sample signal to assist in obtaining the genuine absorbance values of the analytes. The materials and circumstances utilized in UV-visible spectroscopy investigations must be understood. The majority of plastic cuvettes, for example, are unsuitable for UV absorption investigations because plastic absorbs UV radiation. Glass can operate as a UVC (100-280 nm^2) and UVB (280-315 nm^2)filter, absorbing the bulk of UVC and UVB while allowing some UVA (315-400 nm^2) to pass through. Because quartz is transparent to the majority of UV light, quartz sample holders are required for UV testing. Because molecular oxygen in the air absorbs wavelengths of light shorter than around 200 nm, it can be thought of as a filter. For observations with wavelengths less than 200 nm, a unique and more expensive setup is required, which normally involves an optical system filled with pure argon gas. There are also cuvette-free systems that allow for the analysis of very small sample quantities, such as in DNA or RNA assays.
- Detectors: After the light has gone through the sample, a detector converts the light into an electronic signal that can be read. Detectors are often made of photoelectric coatings

or semiconductors. When exposed to light, a photoelectric coating emits negatively charged electrons. An electric current proportionate to the light intensity is generated when electrons are expelled. In UV-visible spectroscopy, a photomultiplier tube (PMT) is one of the most used detectors. A PMT uses the photoelectric effect to expel electrons when exposed to light, then multiplying the emitted electrons sequentially to create a greater electric current. PMT detectors are very helpful for detecting extremely low light levels. An electric current proportionate to the light intensity can travel through semiconductors when they are exposed to light. More precisely, photodiodes and charge-coupled devices (CCDs) are two of the most prevalent semiconductor detectors. The signal is identified and sent to a computer or screen once the electric current is created by whatever detector was employed. A graph showing absorbance, optical density, or transmittance as a function of wavelength can be used to illustrate UV-visible spectroscopy data. The wavelength corresponding to the target substance's maximal absorption is chosen for investigation in UV-visible spectroscopy. Because the biggest reaction is produced for a given analyte concentration, this option assures maximum sensitivity. The food green dye has two maximum absorbance peaks: a lesser maximum absorbance peak at 435 nm and a more intense maximum absorbance peak at 619 nm. The highest absorbance peak at 619 nm was utilized for analysis to attain maximum sensitivity when computing an unknown concentration of food green. Diluting a stock solution, taking absorbance measurements, and displaying these on an absorbance versus concentration graph to develop a numerical relationship between concentration and absorbance were used to create standard solutions across a range of known values. A least squares linear regression equation was used to produce a calibration curve. The better the fit, the closer the data points are to a straight line. There is no such thing as a flawless procedure, and UV-visible spectroscopy is no exception. The approach does, however, offer a few key advantages that make it popular, which are stated here. The method is nondestructive, allowing the sample to be reused or processed or analyzed further. Measurements may be taken fast, making them straightforward to include into experimental techniques. Instruments are simple to use and need little user training before usage. In most cases, data analysis necessitates minimum processing, implying that little user training is necessary. The device is relatively affordable to purchase and run, making it suitable for a wide range of laboratories.

3.6.1.6 Fourier-Transform Infrared Spectroscopy (FTIR)

Absorption spectroscopy techniques (FTIR, ultraviolet-visible ("UV-vis") spectroscopy, and so on) are used to determine how much light a sample absorbs at different wavelengths. The simplest method, known as "dispersive spectroscopy," involves shining a monochromatic light beam at a sample, measuring how much of the light is absorbed, and repeating the process for each distinct wavelength. The Fourier transform spectroscopy method is a less obvious means of obtaining the same data. Rather than flashing a single wavelength of light at the sample, this approach shines a beam comprising many frequencies of light at the same time and evaluates how much of it is absorbed by the sample. The beam is then altered to incorporate a new frequency combination, yielding a second data point. Over a short period of time, this procedure is repeated several times. After that, a computer uses all this information to calculate absorption at each wavelength by working backwards. Starting with a broadband light source—one that contains the whole range of wavelengths to be measured—the beam described above is created. The light shines through a Michelson interferometer, which is a set of mirrors with one of them being moved by a motor. Due to wave interference, the interferometer periodically blocks, transmits, blocks, transmits each wavelength of light in the beam as it advances. Different wavelengths are modulated at distinct rates, resulting in a different spectrum for each instant or mirror position in the interferometer.

- Infra-red (IR) Sources: The mid and near IR ranges are where FTIR spectrometers are most commonly employed. The most frequent source for the mid-IR region, 225 m (5,000–400 cm^{-1}), is a silicon carbide element heated to roughly 1,200 K (Globar). The output is comparable to that of a blackbody. Shorter near-IR wavelengths, such as 12.5 m (10,000–4,000 cm^{-1}), need a higher temperature source, such as a tungsten-halogen lamp. Due to the absorption of the quartz envelope, the long wavelength output of these is restricted to roughly 5 m (2,000 cm^{-1}). A mercury discharge lamp produces more power in the far-IR, especially at wavelengths greater than 50 m (200 cm^{-1}), than a thermal source.

- Detector: Pyroelectric detectors, which respond to variations in temperature as the intensity of IR radiation falling on them fluctuates, are extensively used in near-IR spectrometers. Deuterated triglycine sulfate (DTGS) or lithium tantalate are the sensitive elements in these detectors ($LiTaO_3$). These detectors work at room temperature and have enough sensitivity for most everyday applications. A scan normally takes a few seconds to attain the greatest sensitivity. For circumstances demanding better sensitivity or faster reaction, cooled photoelectric detectors are used. In the mid-IR, liquid nitrogen-cooled mercury cadmium telluride (MCT) detectors are the most common. An interferogram may be detected in as little as 10 ms with these detectors. Near-IR systems often use uncooled indium gallium arsenide photodiodes or DTGS.
- Beam splitter: In a perfect beam splitter, half of the incident light is transmitted, and half is reflected. Due to the restricted range of optical transmittance of any material, many beam splitters can be employed interchangeably to span a large spectrum range. The beam splitter for the mid-IR region is commonly composed of KBr with a semi-reflective germanium-based coating. Because KBr absorbs heavily beyond 25 m (400 cm^{-1}), CsI or KRS-5 are occasionally used to increase the wavelength range to around 50 m (200 cm^{-1}). Where moisture vapor is a concern, ZnSe is an option that is restricted to roughly 20 m (500 cm^{-1}). CaF_2 is the most common near-IR material, as it is both tougher and less susceptible to moisture than KBr, however, it can only be utilized up to around 8 m (1,200 cm^{-1}). One beam goes through the beam splitter twice in a basic Michelson interferometer, while the other passes through just once. An extra compensator plate of identical thickness is included to compensate for this. The majority of far-IR beam splitters are made of polymer sheets and have a limited wavelength range.

FTIR can be employed in any application that previously required the use of a dispersive spectrometer (see external links). Furthermore, the increased sensitivity and speed have opened up new application possibilities [34]. Spectra may be detected even when only a little amount of energy enters the detector, and scan speeds can approach 50 spectra/s. In geology, chemistry, materials science, and biology, Fourier transform infrared spectroscopy is employed. In hydrophobic membrane settings, FTIR is also employed to analyze different nanomaterials and proteins. The capacity of FTIR to directly assess the polarity at a certain point along the backbone of a transmembrane protein has been demonstrated in studies. The bonding characteristics of diverse organic and inorganic nanomaterials, as well as their quantitative investigation, may be done using FTIR.

3.6.1.7 Raman Spectroscopy

Raman spectroscopy relies on Raman scattering, which is the inelastic scattering of photons. A monochromatic light source is employed, which is commonly a laser in the visible, near infrared, or near ultraviolet spectrum, however, X-rays can also be used. The energy of the laser photons is pushed up or down when the laser light interacts with molecular vibrations, phonons, or other excitations in the system. The energy shift reveals information about the system's vibrational modes. Infrared spectroscopy usually produces comparable but complimentary results. A laser beam is usually used to light a sample. Typically, a sample is illuminated using a laser beam. Electromagnetic radiation from the illuminated area is gathered by a lens and directed through a monochromator. Elastic scattered radiation at the laser line's wavelength (Rayleigh scattering) is filtered out by a notch filter, edge pass filter, or band pass filter. The remaining light is then dispersed onto a detector filters off elastic scattered radiation at the wavelength corresponding to the laser line (Rayleigh scattering), while the remainder of the collected light is spread onto a detector. A laser beam is frequently used to light a sample. A lens collects the electromagnetic radiation from the lit area and sends it through a monochromator. The rest of the collected light is distributed onto a detector after elastic scattered radiation at the wavelength corresponding to the laser line (Rayleigh scattering) is filtered out by a notch filter, edge pass filter, or band pass filter. Because spontaneous Raman scattering is often relatively weak, distinguishing the weak inelastically scattered light from the powerful Rayleigh scattered laser light has been a major challenge in collecting Raman spectra for many years (referred to as "laser rejection"). To obtain a high level of laser rejection in the past, Raman spectrometers employed holographic gratings and numerous dispersion stages. Photomultipliers were once the preferred detectors for dispersive Raman setups, resulting in lengthy acquisition times. Modern equipment, on the other hand, nearly always uses notch or edge filters for laser rejection. The most popular spectrographs for use with

NIR lasers are dispersive single-stage spectrographs (axial transmissive (AT) or Czerny–Turner (CT) monochromators) linked with CCD detectors, however, Fourier transform (FT) spectrometers are also widespread [31].

The term "Raman spectroscopy" usually refers to vibrational Raman spectroscopy that uses laser wavelengths that are not absorbed by the material. Surface-enhanced Raman, resonance Raman, tip-enhanced Raman, polarized Raman, stimulated Raman, transmission Raman, spatially-offset Raman, and hyper Raman are all examples of Raman spectroscopy. For tissue imaging using surface-enhanced Raman scattering (SERS)-labeled antibodies, Raman reporter molecules with olefin or alkyne functionalities are being produced. Raman spectroscopy has also been utilized to do real-time, *in situ* biochemical evaluation of wounds using a noninvasive approach. The establishment of a quantitative metric for wound healing process has been made possible by multivariate analysis of Raman spectra. Spatially offset Raman spectroscopy (SORS), which is less sensitive to surface layers than traditional Raman, can be used to detect counterfeit pharmaceuticals without having to open the package and to examine biological tissue non-invasively.

3.7 Conclusion

This chapter describes the nanotechnology used in the synthesis of nanolubricants. These nanolubricants are made from various nanomaterials. Due to increased interest in the nanotechnology, the study of various nanomaterials has been introduced for future perspectives. Their unique physical properties and chemical properties have made them useful for different purposes. The characteristics of these nano lubricants have been studied using different mechanical techniques like UV-vis, Raman spectroscopy and AFM.

References

1. Brenner, D., Mahbooba, Z., Saberi-Movahed, F., Krim, J., Liu, Z., Ivanov, M.G., Shenderova, O., Nanodiamond-based nanolubricants: Experiment and modeling. *MRS Online Proceedings Library (OPL)*, vol. 1703, 2014.
2. Alves, S.M., Mello, V.S., Faria, E.A., Camargo, A.P.P., Nanolubricants developed from tiny CuO nanoparticles. *Tribol. Int.*, 100, 263–271, 2016.

3. Ba, Z., Han, Y., Qiao, D., Feng, D., Huang, G., Composite nanoparticles based on hydrotalcite as high performance lubricant additives. *Ind. Eng. Chem. Res.*, 57, 45, 15225–15233, 2018.
4. Brenner, D., Mahbooba, Z., Saberi-Movahed, F., Krim, J., Liu, Z., Ivanov, M.G., Shenderova, O., Nanodiamond-based nanolubricants: Experiment and modeling. *MRS Online Proceedings Library (OPL)*, vol. 1703, 2014.
5. Paul, G., Hirani, H., Kuila, T., Murmu, N.C., Nanolubricants dispersed with graphene and its derivatives: An assessment and review of the tribological performance. *Nanoscale*, 11, 8, 3458–3483, 2019.
6. Ghazi, S.S., Mezher, M.T., Namer, N.S., Shakir, R.A., Tribological behaviour of AA1060 aluminium alloy with the aid of using nano lubricant additives. *J. Mech. Eng. Res. Dev.*, 44, 296–304, 2021.
7. Shahnazar, S., Bagheri, S., Abd Hamid, S.B., Enhancing lubricant properties by nanoparticle additives. *Int. J. Hydrog. Energy*, 41, 4, 3153–3170, 2016.
8. Shah, R., Woydt, M., Zhang, S., The economic and environmental significance of sustainable lubricants. *Lubricants*, 9, 2, 21, Feb. 22, 2021.
9. Krishnamurthy, A., Razak, K.A., Halemani, B.S., Buradi, A., Afzal, A., Saleel, C.A., Performance enhancement in tribological properties of lubricants by dispersing TiO2 nanoparticles. *Mater. Today: Proc.*, 47, 6180–6184, 2021.
10. Cortes, V., Sanchez, K., Gonzalez, R., Alcoutlabi, M., Ortega, J.A., The performance of SiO2 and TiO2 nanoparticles as lubricant additives in sunflower oil. *Lubricants*, 8, 1, 10, Jan. 15, 2020.
11. Wu, H., Zhao, J., Cheng, X., Xia, W., He, A., Yun, J.H., Jiang, Z., Friction and wear characteristics of TiO2 nano-additive water-based lubricant on ferritic stainless steel. *Tribol. Int.*, 117, 24–38, 2018.
12. Mobarak, H.M., Mohamad, E.N., Masjuki, H.H., Kalam, M.A., Al Mahmud, K.A.H., Habibullah, M., Ashraful, A.M., The prospects of biolubricants as alternatives in automotive applications. *Renew. Sust. Energ. Rev.*, 33, 34–43, 2014.
13. Luo, T., Wei, X., Huang, X., Huang, L., Yang, F., Tribological properties of Al2O3 nanoparticles as lubricating oil additives. *Ceram. Int.*, 40, 5, 7143–9, Jun. 1, 2014.
14. Dandan, A. and Samion, S., Palm oil as an alternative lubricant oil: A brief review. *Colloquium, UTM*, vol. 11, pp. 17–19, 2017.
15. Fedele, L., Colla, L., Minetto, S., Scattolini, M., Bellomare, F., Bobbo, S., Zin, V., Nanofluids characterization and application as nanolubricants in heat pump systems. *Sci. Technol. Built Environ.*, 21, 5, 621–630, 2015.
16. Pena-Paras, L., Maldonado-Cortes, D., Castillo, F., Leal, J., Garza, S., Application of nanoclay lubricants for lowering wear of tools for steel meshing-A case study, in: *IOP Conference Series: Materials Science and Engineering*, vol. 400, IOP Publishing, p. 072004, August 2018.
17. Wu, H., Zhao, J., Luo, L., Huang, S., Wang, L., Zhang, S., Jiang, Z., Performance evaluation and lubrication mechanism of water-based nano-lubricants containing nano-TiO2 in hot steel rolling. *Lubricants*, 6, 3, 57, 2018.

18. Kotia, A., Ghosh, G.K., Srivastava, I., Deval, P., Ghosh, S.K., Mechanism for improvement of friction/wear by using Al2O3 and SiO2/ Gear oil nanolubricants. *J. Alloys Compd.*, 782, 592–599, 2019.
19. Nallusamy, S., Synthesis and performance characterization of nano-lubricants proposed for heavy load ball bearing. *J. Nano Res. Trans. Tech Publications Ltd*, 54, 75–87, 2018.
20. Xia, W., Zhao, J., Wu, H., Zhao, X., Zhang, X., Xu, J., Jiang, Z., Effects of oil-in-water based nanolubricant containing TiO2 nanoparticles in hot rolling of 304 stainless steel. *J. Mater. Process. Technol.*, 262, 149–156, 2018.
21. Nallusamy, S. and Logeshwaran, J., Experimental analysis on nano-lubricants used in multi cylinder petrol engine with copper oxide as nano-particle. *Rasayan J. Chem.*, 10, 3, 1050–1055, 2017.
22. Chandrabhan, S.R., Jayan, V., Parihar, S.S., Ramaprabhu, S., Development of a nitrogen-doped 2D material for tribological applications in the boundary-lubrication regime. *Beilstein J. Nanotechnol.*, 8, 1, 1476–1483, 2017.
23. Sharif, M.Z., Azmi, W.H., Redhwan, A.A.M., Zawawi, N.M.M., Preparation and stability of silicone dioxide dispersed in polyalkylene glycol based nanolubricants, in: *MATEC Web of Conferences*, vol. 90, EDP Sciences, p. 01049, 2017.
24. Azman, N.F., Syahrullail, S., Rahim, E.A., Preparation and dispersion stability of graphite nanoparticles in palm oil. *J. Tribol.*, 19, 132–141, 2018.
25. Diabb, J., Rodríguez, C.A., Mamidi, N., Sandoval, J.A., Taha-Tijerina, J., Martínez-Romero, O., Elías-Zúñiga, A., Study of lubrication and wear in single point incremental sheet forming (SPIF) process using vegetable oil nanolubricants. *Wear*, 376, 777–785, 2017.
26. Ba, Z., Han, Y., Qiao, D., Feng, D., Huang, G., Composite nanoparticles based on hydrotalcite as high performance lubricant additives. *Ind. Eng. Chem. Res.*, 57, 45, 15225–15233, 2018.
27. Alves, S.M., Mello, V.S., Faria, E.A., Camargo, A.P.P., Nanolubricants developed from tiny CuO nanoparticles. *Tribol. Int.*, 100, 263–271, 2016.
28. Gupta, H., Rai, S.K., Kuchhal, P., Anand, G., Characterization and experimental investigation of rheological behavior of oxide nanolubricants. *Part. Sci. Technol.*, 42, 4, 622–632, 2020.
29. Zeinali Heris, S., Razbani, M.A., Estellé, P., Mahian, O., Rheological behavior of zinc-oxide nanolubricants. *J. Dispers. Sci. Technol.*, 36, 8, 1073–1079, 2015.
30. Dai, W., Kheireddin, B., Gao, H., Liang, H., Roles of nanoparticles in oil lubrication. *Tribol. Int.*, 102, 88–98, 2016.
31. Tosi, M.M., Ramos, A.P., Esposto, B.S., Jafari, S.M., Dynamic light scattering (DLS) of nanoencapsulated food ingredients, in: *Characterization of Nanoencapsulated Food Ingredients*, pp. 191–211, Academic Press, Elsevier.com/academic-press, 2020.
32. Lim, J., Yeap, S.P., Che, H.X., Low, S.C., Characterization of magnetic nanoparticle by dynamic light scattering. *Nanoscale Res. Lett.*, 8, 1, 1–14, 2013.

33. Theivasanthi, T. and Alagar, M., Titanium dioxide (TiO_2) nanoparticles XRD analyses: An insight. arXiv preprint arXiv:1307.1091, 2013.
34. Latif, M.S. *et al.*, Effect of temperature on the synthesis of Centella asiatica flavonoids extract-mediated gold nanoparticles: UV-visible spectra analyses. *AIP Conference Proceedings*, vol. 2016, AIP Publishing LLC, 2018.

4

Metal-Based Nanolubricants: Current and Future Perspectives

Deepak Gupta[1]*, Chandra Kumar[2], Aakash Mathur[1], Shruti Mishra[1], Anis Ahmad[1], Namrata Deka[3], Priyanki Kalita[4] and Milan Singh[1]

[1]*School of Basic and Applied Sciences, Galgotias University, Greater Noida, Uttar Pradesh, India*
[2]*Departamento de Física, Facultad de Ciencias, Universidad Católica del Norte, Antofagasta, Chile*
[3]*Polymer Research Centre, Department of Chemical Sciences, Indian Institute of Science Education and Research, Kolkata, West Bengal, India*
[4]*Department of Physics, National Institute of Technology Meghalaya, Bijni Complex, Shillong, Meghalaya, India*

Abstract

One of the best methods to reduce friction and wear is by using nanoadditives in lubricants, which is crucial for preserving energy, cutting emissions, and protecting the environment. Nanolubricant additives have made significant strides in both scientific research and industrial applications with the advancement of science and technology. In comparison to organic molecules typically utilized as lubricant additives, nanoparticles provide a number of significant advantages, including their nanometer size, which enables them to access the contact area readily. This review focuses on how tribological performance in oil lubrication is affected by metal-containing nanoparticles. The review also includes a thorough discussion of the additives' dispersion stabilities in lubricants. Analyses are done on how the addition of nanoparticles affects the various performance characteristics. The review also outlines the lubrication mechanisms of nanolubricant additives, such as tribofilm production, microbearing effect, self-repair performance, and synergistic impact. Analysis is done on the effects of nanoparticle size, shape, surface functionalization, and concentration on wear and friction. Consideration is given to the challenges and potential of developing metal-containing nanoparticles as lubricant additives. Fundamental information on metal, metal oxides,

**Corresponding author*: dgupta.nitm@gmail.com

Mohd Yusuf, Lalit Prasad and Shafat Ahmad Khan (eds.) Nanolubricants: Generation and Applications, (89–134) © 2024 Scrivener Publishing LLC

metal sulfides, and rare earth nanoparticle-based nanolubricants is generalized. The difficulties and potential of nanolubricant additives are also discussed, which directs the creation of innovative additives with important lubrication and anti-wear capabilities in the future.

Keywords: Friction, anti-wear, nanoadditives, lubricants, metal applications

4.1 Introduction

Lubrication is a procedure for reducing wear and friction on two adjacent moving surfaces [1]. The advantages of lubrication include the prevention of rust, water, and dust, as well as acting as an insulator in transformers. Wear and friction can result in energy losses and machinery breakdown (for instance, in engines, shafts, bearings, and gears). Therefore, having enough lubrication is crucial to resolving these problems [2]. Mineral oil has been used as a lubricant for a very long time. However, environmental damage brought on by the dumping of mineral oil directly impacted aquatic and terrestrial ecosystems [3]. A viable substitute for mineral oil is an alternate source, such as biolubricant oil, because it is harmless and biodegradable. Small weight percentages of additives are added to the lubricant base stock to improve and enhance the oil's characteristics, which helps to further reduce wear and friction in the system [2]. These additives might be dispersant, anti-corrosion, anti-wear (AW), and extreme pressure (EP) additives. AW and EP additives are particularly utilized to lubricate mechanical parts. Traditional EP and AW additives consist of chlorine and phosphorus compounds, but their usage has been constrained for environmental reasons. Researchers are looking into the design of nanoparticles (NPs) as a new class of lubricant additives, as a means of resolving this issue and the performance bottlenecks that have been reached by the majority of currently available lubricants [4]. The dispersion of NPs in lubricating fluids is known as nanolubrication. The tribological and thermophysical characteristics of base oils are improved by NPs with sizes less than 100 nm due to their superior optical, mechanical, and electrical qualities [5]. Table 4.1 lists the base oils being categorized by the American Petroleum Institute. Excellent wear and friction reduction can be found in metal [6], metal sulfides [7], metal oxide [8], and rare earth compounds [9]. In the current chapter, a thorough literature review is used to describe the properties and probable role of mainly metal-based nanolubricants.

Table 4.1 Base oil categorized by the American Petroleum Institute [51–53].

	Base oil category	**Sulfur %**		**Saturates %**	**Viscosity index**	**Temperature range**	**Market value**	**Manufacturing process**
Mineral	Group 1	>0.03	And/or	<90	80 to 120	32-150°F	cheapest base oils	Simple refining
	Group 2	<0.03	And	>90	80 to 120	-	-	Hydrocracking
	Group 3	<0.03	And	>90	> 120	-	-	Severely hydrocracked (higher pressure and heat)
Synthetic	Group 4	PAO synthetic lubricants				Much broader temperature range		
	Group 5	All oils excluded in group 1to 4. Ex silicone, phosphate ester, polyalkylene glycol (PAG), polyolester, biolubes, etc.						

4.2 Synthesis Mechanism of NPs

There is a wide variety of synthetic techniques that have been developed to synthesize metal NPs, and these techniques can be broadly classified into two categories, top-down methods and bottom-up methods.

4.2.1 Top-Down Methods

The top-down methods of nanomaterial synthesis involve a bulk starting material being broken down or converted into smaller-sized particles. The size reduction of the bulk material is accomplished by various physical or chemical methods. Some of the synthetic techniques categorized as top-down methods, include mechanical milling, sputtering, pulsed laser deposition, and electrospinning, which are briefly discussed below:

4.2.1.1 Ball Milling

Ball milling is a type of mechanical milling process where large-sized particles are reduced in size by a mechanical grinding process using several metal balls. It has been considered as an effective and economical technique for creating various metal NPs and nanocomposite blends. High energy mechanical ball milling has been used to create different aluminum-based metal nanoalloys and nanocomposites. Blending aluminum with other elements, such as magnesium, nickel, and carbon through ball milling, can be seen as an effective way to alter the combustion and thermal properties of the primary material [10]. Although ball milling techniques can successfully synthesize high purity NPs at an industrial scale., this technique shows limitations in aspects of the time required and the energy consumed during the synthetic process.

4.2.1.2 Electrospinning

Electrospinning is a technique used to produce a web of micro/nanofibers without being woven. Metal oxide nanofibers are often synthesized via electrospinning owing to the excellent morphological control afforded by this technique. Figure 4.1 displays a schematic representation of the electrospinning setup, which generally consists of three essential parts: (i) a syringe pump or spinneret; (ii) a high voltage power source; and (iii) a collector. The hypodermic syringe is filled with the mother solution and connected to a spinneret to get the fibers using the electrospinning process.

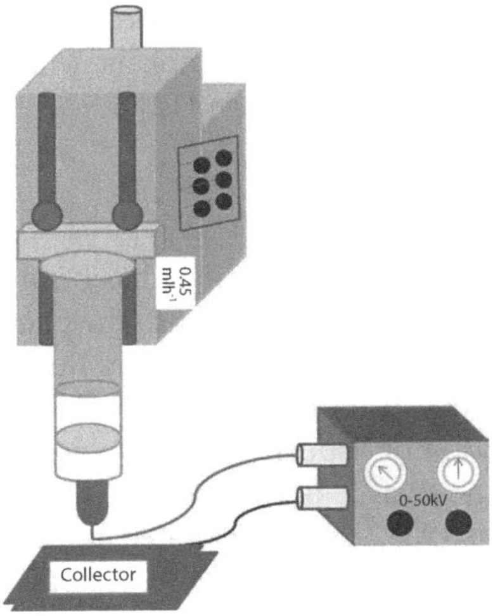

Figure 4.1 Schematic representation of the electrospinning setup.

The syringe is then loaded into the syringe pump, which starts pumping the solution at a constant flow rate. The distance between the syringe and collector must be fixed during the process, and an electric field is applied between the spinneret and the collector. The surface tension of the liquid drop is balanced by the repulsive forces between the surface charges distributed over the polymer solution. When the repulsive force grows larger than the surface tension, the pendant drop transforms into a conical shape called the Taylor cone and exhibits a half angle of 49.3°. The pendant drop's tip eventually releases an electric jet towards the collector and stretches into continuous, nanodimensional fibers [11]. The properties and physical characteristics of the nanofibers can be modified by altering the viscosity of the liquid, surface tension, electric field, etc.

4.2.1.3 Lithography

Lithography is a widely used top-down process for synthesizing various nanoarchitectures. Lithography techniques employ a three-dimensional (3D) surface mold that is embedded onto a heated cast. High resolution surface topology in nanomaterials can be obtained through nanolithography. There are several examples of lithographic techniques used in metal-based nanomaterial synthesis, such as electron beam lithography (EBL), nanosphere lithography (NSL), nanoimprint lithography, and hole-mask lithography. These lithography techniques fall into either of the two categories—masked lithography and maskless lithography. Maskless lithographic techniques result in arbitrary patterns, whereas masked lithography results in a uniform surface pattern [12].

4.2.1.4 Sputter Deposition

Sputter deposition or sputtering technique has been well-explored for the generation of thin films of nanodimensions. The phenomenon of thin film growth by sputtering was initially observed by Bunsen and Groves in the early 1800s. Sputter deposition involves the irradiation of a charged species and the acceleration of vaporized atoms towards a substrate to generate a thin film on the surface of the substrate. The bombarding ions are usually noble gas ions (Ar^+, He^+) accelerated by an applied electric field. The removed species or the sputtered species from the target surface are extremely energetic atoms that reach the substrate with energies ranging from 1 to 10 eV as shown in Figure 4.2. One of the dominant sputtering source for thin film deposition is the magnetron source [13].

Various nanodimensional thin layers of high purity metals, semiconductors, doped semiconductors, etc., have been synthesized at comparatively low costs with the help of magnetron sputtering technique. On the other hand, nanofilms of metal chalcogenides (metal oxides, metal sulfides, etc.), metal alloys, metal-based nanocomposites are synthesized using techniques such as reactive sputtering or co-sputtering. In the co-sputtering

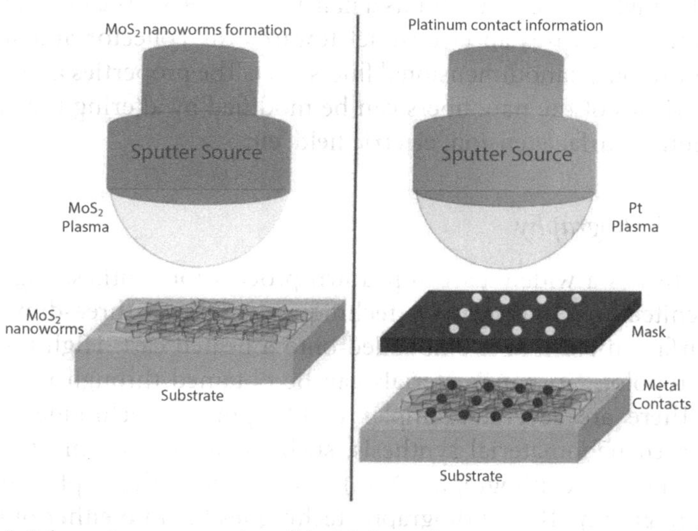

Figure 4.2 Schematic representation of the MoS_2 synthesis using sputtering technique.

process, two neighboring targets may be sputtered such that materials from both targets are incident on the substrate, depositing a thin film of the composite or compound. The composition of the films can be precisely controlled by changing the working pressure, power, and the ratio of the inert gas to reactive gases [13].

4.2.1.5 Pulsed Laser Deposition

Pulsed laser deposition (PLD) is another flexible method for creating thin films by perforating one or more targets with the help of concentrated laser pulses. This method has been employed extensively for the deposition of oxides, nitrides, carbides, metallic, and polymer with controlled chemical compositions. This technique employs a high-power ultraviolet pulsed laser to irradiate the target. Figure 4.3 depicts the design of the experimental setup of a typical PLD method to deposit thin films. The pulsed laser is passed through a quartz lens to tune the intensity of the laser. The laser pulse strikes the target of desired composition at an angle of 45°. The total energy output of the laser beam can be controlled with the mask located at

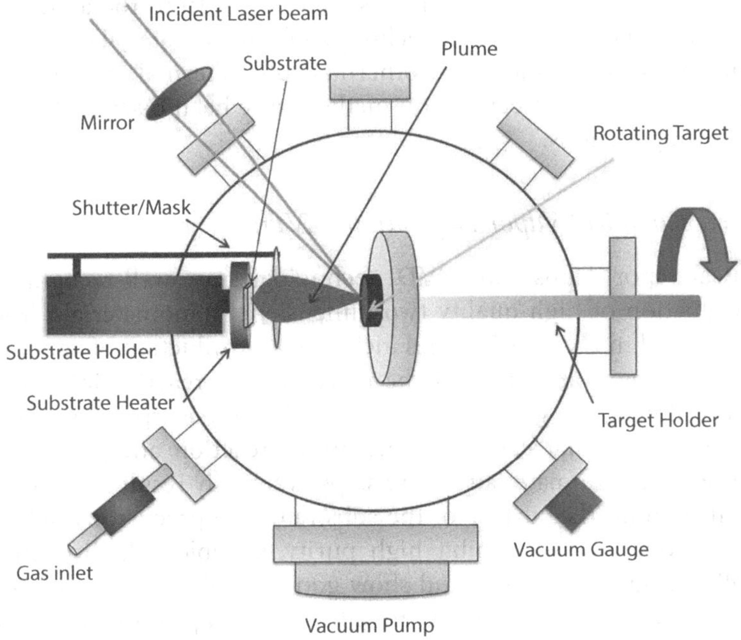

Figure 4.3 Schematic illustration of a pulsed laser deposition (PLD) setup.

the output window of the laser source. Further, the target is mounted on a rotating carousel, which can accommodate up to four targets. Vaporization of the absorption depth starts within the first 100 ps of a 25-ns pulse. After this, the target material under the plasma will be screened from the remainder of the laser pulse, which will be absorbed by the plasma, as it becomes more directional. The congruent ablation in PLD process leads to simultaneous ejection of the components of target material as the laser energy is absorbed in a small volume (~10–13 m^3). This ensures the stoichiometric transfer of target material onto the surface of the substrate. The ejection of the target material is followed by the formation of plasma or plume, which expands very fast in a perpendicular direction to the plane of the target. The ejected target material is deposited on the substrate, which is placed on a heated substrate holder. Some of the major benefits of the PLD technique are the accurate transfer of stoichiometry between the target and the thin film that is being deposited and the rapid deposition rate (0.1 nm per pulse).

4.2.2 Bottom-Up Approaches

Bottom-up approaches for nanomaterial synthesis utilize different physical or chemical techniques to assemble smaller units from the atomic scale. Various well-known bottom-up techniques for nanomaterial synthesis like chemical vapor deposition, hydrothermal/solvo-thermal method, sol–gel synthesis, and the co-precipitation method have been discussed briefly in the following sections.

4.2.2.1 Chemical Vapor Deposition (CVD)

Chemical vapor deposition (CVD) technologies are well-recognized for the production of high-quality two-dimensional nanomaterials. The key steps involved in CVD techniques include (i) introduction of vapor in the reactor and the evaporation of the reactants; (ii) reactive intermediates produced during the gas-phase reactions; (iii) physisorption of the reactive intermediates that leads to further surface reaction; and (iv) nucleation and growth of nanomaterials. These steps eventually result in the creation of a thin film on the surface of the substrate. The precursor used in this synthesis route should exhibit high purity, suitable volatility, environmentally benign conditions, and show good stability during evaporation [14]. Moreover, it is worth mentioning that the composition used in CVD should not result in residual impurities.

4.2.2.2 Hydrothermal/Solvothermal Methods

The hydrothermal process is the most popular approach among materials science researchers for synthesizing nanomaterials. Hydrothermal/solvothermal methods can be applied over a wide range of temperature, irrespective of the solvent in the system. This synthetic approach is known as hydrothermal if the reaction is in aqueous media, and solvothermal if the reaction is in an organic solvent. The morphologies of the NPs can be altered by tuning the vapor pressure of the reacting composition [14]. The starting components of the reaction mixture are transferred into a sealed autoclave composed of stainless steel coated with Teflon and heated in an oven over the solvent's boiling point. This causes a significant increase in the vapor pressure within the reaction autoclave (as schematically displayed in Figure 4.4). The high temperature and high pressure combination synergistically produces highly crystalline materials in a single step and removes the need for post-annealing treatments. There have been numerous reports of a wide array of morphologies in the metal NPs synthesized through the hydrothermal methods. Overall, the properties of the NPs synthesized through this method is highly influenced by the concentration of reactants, stabilizing agents, reaction time, and pressure conditions.

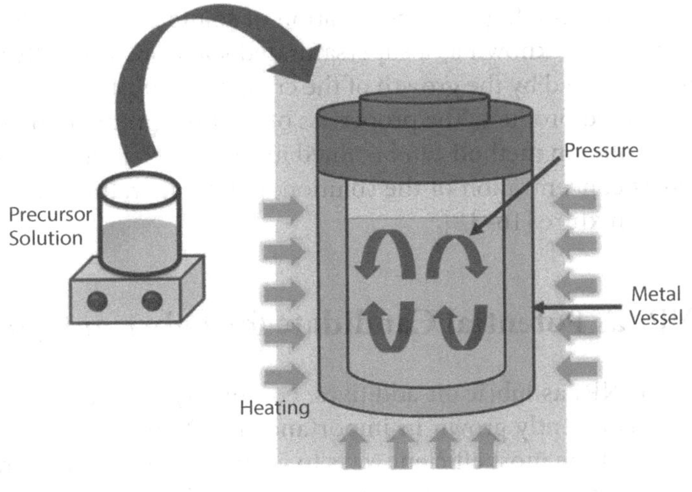

Figure 4.4 Schematic diagram of the hydrothermal method.

4.2.2.3 Sol–Gel Method

The sol–gel synthetic technique is a form of wet-chemical synthetic process commonly used in the synthesis of metal oxide NPs. The sol–gel method is preferably used in material fabrication because this method provides optimal particle size material with phase purity. The reactive precursor involved in this technique was dissolved in a solvent and converted to a gel form through the application of heat. This process generally involves high synthesis temperatures between 600°C and 800°C. This technique consists of hydrolysis of the precursor solution, followed by a condensation reaction. Chelating agents such as polyethylene–poly (ethylene glycol), citric acid, and polyvinyl alcohol (PVA) are generally employed to aid the homogenization of the reaction media [15].

4.2.2.4 Co-Precipitation

The solution co-precipitation method is another well-known wet-chemical synthetic method, which is usually based on the separation and collection of specific elements from the solution in the form of their respective insoluble compounds such as hydroxides and carbonates [16, 17]. During co-precipitation, the following phenomena would occur: (1) mixing of liquid or super-saturation; (2) nucleation; (3) the formation of primary products due to crystal growth; and (4) their agglomeration. In the co-precipitation method, the specific ions in the solution and their corresponding counter ions exceed the solubility of the end product. Such a solution, which has a higher concentration of solute than the solubility of the end product, is known as a supersaturated solution. Thereafter, nucleation starts followed by the growth of the crystalline substance, and results in a well-ordered product. The properties of the final product obtained via the co-precipitation method can be tuned just by controlling some parameters such as concentration of the solution, pH, temperature, and stirring speed of the mixture [16–18].

4.3 NPs as Potential Candidate for Lubricant Additive

The usage of NPs as lubricant additives, commonly referred to as nanolubricants, has recently grown in importance [4]. Nanoadditives in lubricants are one of the most efficient ways to reduce friction and wear, which is crucial for energy conservation, emission reduction, and environmental protection. Nanolubricant additives have made significant strides in both scientific research and industrial applications with the advancement of science and technology. Due to their small diameters, nanomaterials can easily infiltrate areas of friction contact and can create a protective tribofilm, avoiding wear on the surface of friction pairs. Nanomaterials also have high surface activity, which can enhance the protective tribofilm's film-forming stability through physical and/or chemical adsorption effects [19].

4.3.1 Nanometal-Based Lubricant Additives

Pure metals, metal oxides, metal sulfides, metal hydroxides, and metal salts are the main components of the nanometal-based lubricant additives. For the very first time, in the 1980s–1990s, Cu NPs as lubricant additives were found to display excellent tribological properties in base oils [20]. Recent studies have also proven that the addition of metallic NPs like Fe, Pd and Ni improves tribological characteristics [21–23]. It has been discovered through comparative research that metal oxides with superior lubricating qualities include ZnO, TiO_2, and ZrO_2 [24–26]. The characteristics of nanometal oxides are also shown to be directly related to particle size, hardness, and concentration. In the field of lubrication, metal sulfides as lubricant additives have also garnered a lot of interest. For instance, tungsten disulfide (WS_2) and molybdenum disulfide (MoS_2) can chemically interact with friction pairs to form a persistent adsorption coating on rubbing surfaces, effectively enhancing the tribological properties of lubricants under certain high friction situations [27]. Layered double hydroxides (LDHs), among metal hydroxides, have also been demonstrated to exhibit exceptional lubricating performance. In particular, under boundary lubrication conditions, LDHs powders display outstanding friction and wear reduction capabilities due to the laminated structure, small particle size, and strong chemical activity. $La(OH)_3$ NPs, a different type of metal hydroxide, have also been demonstrated to be efficient lubricant additives, in addition to LDHs [28].

4.3.2 Coefficient of Friction (COF) and Anti-Wear Properties of Nanolubricants

Two distinct types of friction tests were conducted: durability and wear progression tests. Initially, a durability study is conducted to determine the impact of the NP filler on the wear life of the film. The tests are conducted with various normal loads until a significant rise in COF is noticed. A significant increase in the COF indicates that the sample's film has worn away

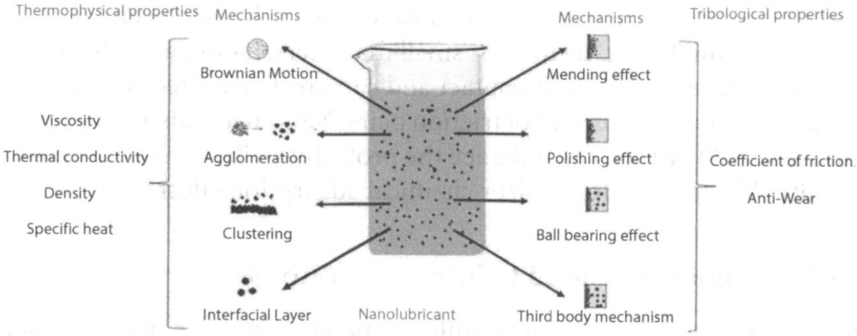

Figure 4.5 Graphical representation of different mechanisms involved with a nanolubricant. Reprinted with permission from Kotia *et al.* [30], Copyright © 2018, Springer-Verlag GmbH Germany, part of Springer Nature.

enough to expose the substrate to its counter surface. Therefore, the number of rubbing cycles up to this point signifies the film's wear life or durability. As the friction test advances, the film undergoes structural changes induced by heat. This causes the friction to gradually increase over time, typically exceeding the usual COF range, until the coating is fully removed [29]. The colloidal effect, small-size effect, protective film effect, rolling (ball-bearing) effect, and the third body effect are the mechanisms through which lubricants containing nanoadditives minimize friction and wear as shown in Figure 4.5.

4.3.3 Lubrication Mechanisms

The NP-mediated improvement in tribological properties of lubricants is governed by various mechanisms. These mechanisms may either have a direct effect of NPs or a secondary effect where NPs are involved for surface enhancement [31]. Various research groups have highlighted the role of different mechanisms for different systems using surface analysis techniques. Some of the important mechanisms include the rolling-effect [8], protective film formation [32], mending effect [33], and polishing effect [34]. These are shown schematically in Figure 4.6.

4.3.4 Rolling Effect or Ball-Bearing Effect

This mechanism is analogous to inclusion of a small ball bearing in between the surface contact area, which reduces the friction coefficient. NPs of spherical morphology act as a ball bearing, which changes the nature of

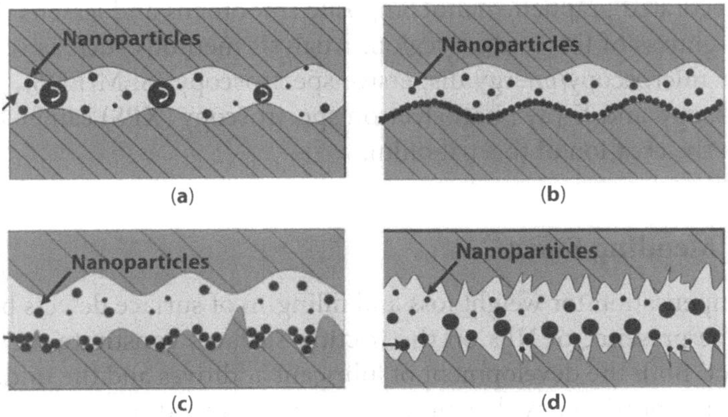

Figure 4.6 (a) Ball rolling mechanism by NPs-based lubricant; (b) protective film formation mechanism by NPs-based lubricant; (c) mending effect mechanism by NPs-based lubricant; and (d) polishing effect mechanism by NPs-based lubricant. Reprinted with permission from Jason *et al.* [2], Copyright © 2020, MDPI.

friction in the contact area from sliding friction to a mixture of sliding and rolling friction [35]. Wu *et al.* have shown the tribological properties of TiO_2 nanolubricants by ball-on-disc tribometer [36]. Here, the TiO_2 NPs in the employed water-based lubricants are supposed to behave as ball bearings between the ball and the disk, facilitating the decrease in friction [36]. Kolodziejczyk *et al.* demonstrated the feasibility of surface-modified Pd NPs in lubricant oils with tremendous tribological properties. Here, the NPs trapped between the contact areas heals the wear tracks by forming a protective layer on the ball's surface, thus accommodating the sliding motion [37].

4.3.5 Protective Film Formation

A protective film layer generally refers to a layer covering the friction surfaces and preventing them from coming into direct contact. This layer formation can be commenced by either NPs getting absorbed in continuous fashion between the contact surfaces, forming a boundary film [38], or friction heat compelling the NPs to melt down resulting in the formation of protective film, which covers the friction surfaces [39]. Also, the NPs are tribo-sintered to make the film, and the compressed and smooth tribofilm (also known as "glaze") is conducive to the reduction of wear and friction [40]. According to Meng *et al.*, a protective coating made of silver-decorated graphene nanocomposite smoothens and lessens the surface roughness of the contact region. Analysis methods such as scanning electron microscopy/energy dispersive spectroscopy (SEM/EDS), Raman spectroscopy, and X-ray photoelectron spectroscopy (XPS) can be used to confirm the creation of this tribofilm.

4.3.6 Mending Effect

The compensation for weight loss and filling-in of surface defects brought on by the presence of NPs on the friction surfaces constitute the mending effect. Both the development of lubricant additives and the underlying theory of nanotribology depend heavily on the mending effect. Liu *et al.* demonstrated the mending effect of Cu NPs using scanning tunnelling microscopy (STM). It was shown that heat generated during friction leads to diffusion of Cu NPs and finally deposition, showing the mending effect [33]. The lubricating oil's NPs have the capacity to fill the friction surface's scars and grooves. Figure 4.6c depicts a schematic of a friction surface with NPs injected into the grooves and scars [41].

4.3.7 Polishing Effect

The polishing effect commits to reduce the roughness of the lubricating surfaces with the assistance of abrasion caused by NPs. The spaces between the rough asperities in tribological contacts got filled with NPs, which may serve as reservoirs for solid lubricants (NPs) inside the contact. The smoothing-out process is the term used to describe the process of filling in abrasive valleys. Because of the reduced surface roughness, this "artificial smoothing" or polishing mechanism improves tribological performance (refer to Figure 4.7). Koshy *et al.* reported this reduced surface roughness using atomic force microscopy (AFM) before and after the deposition of NPs [42].

4.3.8 Surface Modified NP for Nanolubrication

The main obstacle to the widespread commercial use of nanolubricants is the production of stable, homogeneous nanolubricants. NPs possess a high surface area to volume ratio, which results in high surface energy. However, a high surface area indicates high molecular attraction, which causes the particles to clump together (i.e., agglomeration). Agglomeration is the main issue with the stability of the NPs in which they tend to cluster

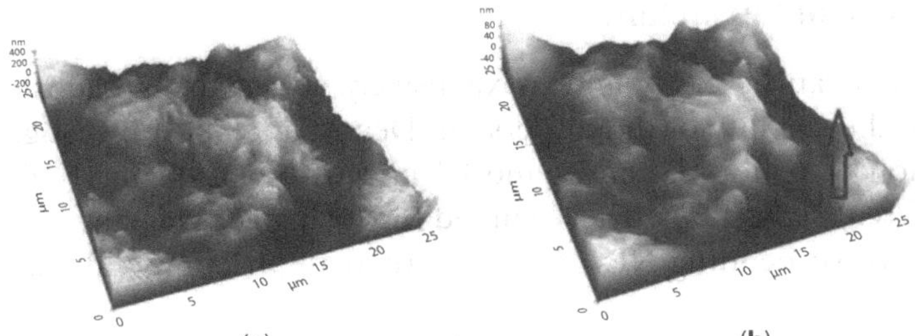

Figure 4.7 (a) AFM images of friction surface before sliding (surface roughness = 143 nm) and (b) AFM images of friction surface after sliding (surface roughness = 68 nm). Reprinted with permission from Koshy *et al.* [42], Copyright © 2015, Elsevier.

as a result of van der Waals forces. It also causes sedimentation, as well as a retarded tribology. As a result, NPs dispersion stability is highly desirable for consistent lubricating performance. The stability of nanolubricants has been greatly improved through physical treatments, the application of surfactants, and the surface modification of NPs.

4.4 Methods to Enhance Dispersion Stability of Nanolubricants

4.4.1 Physical Method

Several mechanical energy techniques, such as mechanical stirring, ultrasonication, ball milling, and high-pressure homogenization, can be used to disperse NPs in base lubricants. It was discovered that ultrasonication was a more effective way to lessen the aggregation of NPs compared to mechanical stirring and shaking. The collapse of unstable cavitation bubbles and a considerable energy loss are caused by ultrasonication, which is caused by a pressure wave that propagates an ultrasonic wave. During the implosive collapse, unstable cavitation bubbles will shatter aggregates. [43, 44]. Mosleh *et al.* [45] observed that molybdenum disulfide (MoS_2), tungsten disulfide (WS_2), and hexagonal boron nitride (hBN) NPs were dispersed in the metal working fluid CIMFLO 20, but are still aggregated after ultrasonication. Therefore, they recommended using a surfactant in the formulation of nanolubricants to improve the dispersion stability.

4.4.2 Use of Surfactant

Surfactants can be easily and inexpensively added to nanolubricants to improve the stability of the dispersion. Demas *et al.* [46] investigated the dispersion of nanoMoS_2 and nanoBN in poly-alpha-olefin (PAO) using five different surfactants. They claimed that using a surfactant was advantageous for suspending NPs, as well as reducing friction and wear on its own.

The process of steric stabilization involves coating of NPs with surfactant molecules, whose long loops and tails extend into solution and create a coating that separates the NPs from other NPs. The presence of a double layer of electric charges around the NPs inhibits their ability to coagulate through a process known as electrostatic stabilization. Depending on the type of surfactant used, a surfactant can either use steric stabilization or a mixture of steric and electrostatic stabilization to prevent the suspension from coagulating. When the surfactant adheres to the surface of the NP in a suitable solvent, it stabilizes the suspension by acting as a physical "cushion" between clashing particles and avoiding NP agglomeration [47]. A water-soluble surfactant is an appropriate surfactant for polar solvents because it can change the hydrophobic surfaces of NPs to hydrophilic surfaces. The appropriate surfactant in a non-polar solvent is an oil-soluble surfactant that modifies the hydrophilic surfaces of the NPs to become hydrophobic [48]. A nonionic surfactant's solubility can be assessed using its hydrophilic-lipophilic balance (HLB) value. The surfactant is more oil-soluble if the HLB value is low, and vice versa. A high HLB surfactant (HLB>10) is suitable for a water-soluble suspension (hydrophilic surfactant), whereas a low HLB surfactant (HLB<8) is suitable for an oil-soluble suspension (lipophilic surfactant) [49]. The HLB value ranges from 1 to 20, and it is possible to use either type of surfactant for a given application. Van der Waals interactions can be avoided by using a nonionic surfactant or polymer to maintain the stability of the suspension by steric stabilization, while ionic surfactants can do so through electro-steric stabilization. To get the best dispersion stability of suspension, the surfactant concentration must be carefully managed [50].

4.4.3 Stability by Modification on Surface

Surface modification techniques for NPs used as lubricant additives have improved substantially in recent years. The surfaces of NPs are chemically modified using chemical agents to improve dispersion stability. Surface modification utilizes organic agents, inorganic agents, silane coupling agents, and polymers to stabilize inorganic NPs in oil-based solutions. As active sites of reactions, NP surfaces must have capping groups in order to be functionalized. The most utilized capped group found on the surface of NPs is the hydroxyl group, which gets absorbed onto the NP's surfaces, preventing agglomeration and coagulation [51]. The organic substance carrying a polar group and a lengthy alkyl chain are used as agents of modifications. This makes it possible for inorganic NPs to dissolve when used in organic solvent [23]. Silane coupling agents have organic functional groups and alkoxy groups, which are used to alter the surfaces of NPs. Surface modifiers help prevent the aggregation of NPs, allowing nanolubricants to be stable for a very long time. In comparison to untreated particles, metallic particles that have undergone chemical treatment show improved dispersion stability. Zhang *et al.* [52] enhanced the stability of NPs of TiO_2 in liquid paraffin containing NPs via utilizing stearic acid (SA) to cover its surfaces. Through Fourier transform infrared spectroscopy results, Zhang *et al.* further validated that SA was successfully grafted on the surface of anatase TiO_2 NPs. Also, by altering the surface of the SiO_2 NPs in liquid paraffin with oleic acid (OA) in ethanol, the stability of the particles was increased.

4.4.4 Metal-Based Nanolubricants

The tribological performance of nanolubricants is strongly affected by the NPs, including metals (Fe, Cu, Ag, Au, Pd), metal oxides (TiO_2, Al_2O_3, ZrO_2, ZnO, CuO), sulfides (CuS), carbon materials (graphene, MWNT, diamond, CNT), and dichalcogenides (WS_2, MoS_2), as well as nanocomposites. The scope of the present chapter is limited to metal-based nanoadditives for improved tribological performances. Metals are among the most prevalent NPs added to lubricants.

Copper (Cu) is relatively softer than other metals; it forms a protective layer on the tribo-pairs, thereby improving the tribological properties of the oil. Cu NPs are widely considered as great anti-wear and extreme pressure agents because they typically have small particle sizes, low melting points, and required ductility. As an additive, Cu NPs considerably enhanced the tribological characteristics of lubricants, enabling essential lubrication. It should be noted that Cu NPs created a tribofilm with low elastic modulus and hardness on the tribo-pairs, thereby improving wear and friction. Their inclusion produced stable nanolubricants and protected them from synthesis-induced degradation [30]. Angela Gondolini *et al.* investigated Cu-based nanolubricants. The generated fluid displayed good stability for at least 60 days post-synthesis with uniformly distributed NPs. At a Cu concentration of 0.01 vol.% at 80°C, these nanolubricants demonstrated a noteworthy performance by lowering the friction coefficient as high as 32.7% [54]. Triangular Cu nanoplates were created by Jing Wang *et al.* as a grease additive with ascorbic acid as the reductant and Cetyltrimethyl ammonium bromide (CTAB) as the capping reagent. These nanoplates created a film at the friction pair's interface during operation, which resulted in an 82.2% reduction in wear loss and considerably enhanced the lubricant's tribological qualities [55]. Matthias Scherge *et al.* studied the Cu^{2+} atoms coated with surfactants and generated a colloidal Cu-based additive for lubricants. This structure resembled a reverse micelle structure, which got uniformly dispersible in the base oil. Their research revealed that the mechanism of Cu tribofilms involved three stages. First, there was an immediate decrease in the COF. Second, there was the formation of a solid lubricant composed of Cu monolayers. Finally, the plastic flow of the Cu tribofilm combined with mechanical intermixing and tribochemical reactions formed a third body. Besides these three phases, the polymerization of monomers present in the base oil resulted in the formation of a protective layer that aids in the transition to hydrodynamic lubrication and inhibited oxidation. These mechanisms provided an efficient reduction in friction and wear. Consequently, the Cu addition provided a long-lasting solution for friction, wear, and prolonged component life [56]. Yang Guang-bin *et al.* used a two-phase extraction method to create Cu NPs surface-modified by dioctylamine dithiocarbamate (DTC8). These NPs had a uniform size distribution and a smaller particle size. The deposition of DTC8+Cu formed a self-repairing protective layer on a worn-out steel surface. DTC8+Cu in liquid paraffin (base oil) demonstrated excellent anti-wear performance. At 1.5 wt% of DTC8+Cu, the wear scar diameter is the smallest [57]. Tribological research revealed that the usage of Cu NPs is the most effective in conjunction with other NPs for reducing friction and wear [21]. **Silver (Ag)** is another softer metallic NP widely used in nanolubricants. Ag NPs increased the tribological properties either by forming a tribofilm or by getting pressed onto the tribo-pairs, thereby, introducing a smoother layer on the friction pairs, which gets sheared easily. With various thiolate ligands, Chanaka Kumara *et al.* created oil-suspendible Ag NPs with sizes ranging from 1-6 nm as shown in Figure 4.8. By adding NPs to the PAO, the boundary lubrication friction and the wear were reduced by up to 35% and 85%, respectively. They claimed that the organic layer's molecular composition on the NP surface had a substantial impact on the friction behavior compared to the size of the NP. Transmission electron microscopy (TEM) revealed that the reductions in friction and wear were due to the development of an Ag-rich 50–100 nm thick tribofilm on worn-out

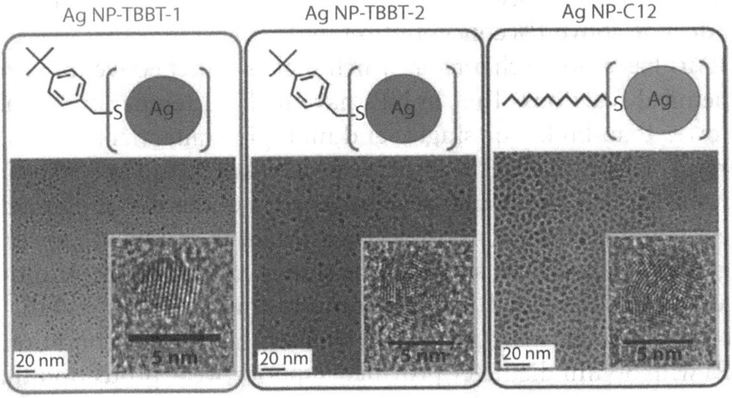

Figure 4.8 Model structures and morphology of thiol-modified Ag NPs. NP-TBBT-1: 4-(tert-butyl)benzyl thiol functionalized, core particle size of 1-3 nm, NP-TBBT-2: 4-(tert-butyl)benzyl thiol functionalized, core particle size of 4-5 nm, NP-C12: Dodecanethiol functionalized, core particle size of 3-6 nm. Reprinted with permission from Kumara *et al.* [58], Copyright (2017) @ ACS Publications.

steel surfaces. Their research revealed that NPs were largely entrapped into the tribofilm by mechanical mixing rather than chemical processes [58]. The extreme pressure and anti-wear characteristics of mineral oil (MO)-based cutting fluids dispersed with Ag NPs were examined by L. Prabu *et al.* in 2018. In the MO-based cutting fluid, the Ag NPs were disseminated in volume fractions of 0.5, 1, and 2 wt percent. Ag NPs reduced the wear scar diameter by 14%. They had a positive impact on MO's anti-wear qualities, increased the cutting fluid's load-wear index by 8%, and decreased the wear scar's diameter by 13%. The concentration of Ag NP in MO caused an increase in the viscosity (at 40°C), flash point, and fire point. At all concentrations, their incorporation improved the nano-cutting fluid's extreme pressure and anti-wear capabilities [59]. By combining Ag NPs at concentrations of 0.1, 0.25, 0.5, and 1 wt% with virgin coconut oil as the base fluid, Muzammil S. Khan *et al.* were able to remove the excess heat during the metal-cutting procedure and lowered the COF. They discovered that adding Ag NPs reduced the COF by 0.25% by facilitating the rolling mechanism over sliding between the steel surfaces. However, above 0.25%, the lubricant transformed from a primarily rolling-based to a sliding-based regime, prompting the friction to rise. The COF dropped for concentrations greater than 0.5% due to higher NP agglomeration. However, surfactant inhibited the agglomeration. The viscosity of such nanofluids rose linearly with the NP concentrations. The cutting power and temperature decreased for the 0.5 wt% concentration, followed by a steady decline above 1% concentration [60].

Gold (Au) has a noble character, which makes it ineffective for corrosion and chemical reactions. The Au NPs have melting temperatures substantially lower than bulk. For single- and multiple-contact conditions, Dave Maharaj *et al.* investigated the impact of Au NPs of 30 nm and 90 nm diameter on the friction forces and wear parameters under dry and liquid environments. In both situations, they claimed that the COF was lower when sliding over a bare silicon substrate, as shown in Figure 4.9. Additionally, surfaces when submerged in the water had lower shear strength. It was found that the COF was comparably smaller than in dry conditions. The Au NPs with 30 nm diameter provided superior wear protection than the

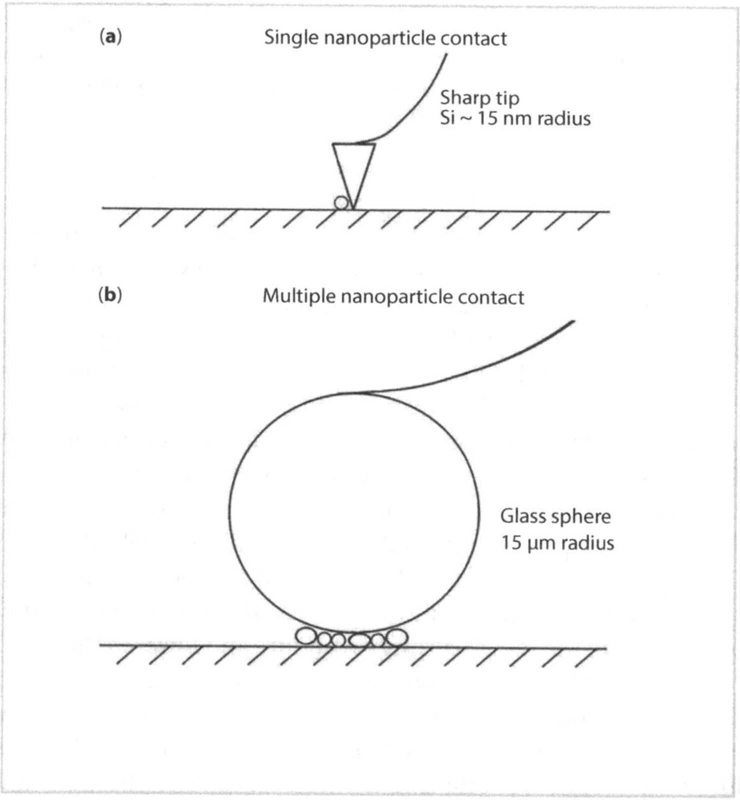

Figure 4.9 Schematics of (a) a sharp tip pushing a particle in a single-particle and (b) a glass sphere sliding over several particles in multiple-particle contact. Reprinted with permission from Maharaj and Bhushan [61], Copyright (2012) @ Beilstein-Institut, under Creative Commons Attribution 2.0 Generic License.

90 nm NPs due to their high particle count [61]. J.C. Sánchez-López *et al.* obtained Au NPs with a mean particle size of 2 nm and a homogenous dispersion via the Brust method. These Au NPs were stabilized via alkanethiol chains by establishing covalent Au–S bonds. A lubricant formulation was prepared by dispersing the Au NPs in paraffin (PAR) oil base at 5 wt%. Tribological studies were performed under intense stress conditions (up to 1.62 GPa). The optical micrographs of the adhered material onto the ball surface for Au NPs and corresponding EDX analysis are presented in Figure 4.10. The results revealed that wear rates fell below 3×10^{-10} mm^3/Nm and the COF remained in the 0.1 range. The Au NPs produced crystallites in the 20–50 nm range due to the friction-induced growth and by detaching the capping layers [23]. Samuel Beckford *et al.* created Au NP-filled PTFE sheets to take advantage of PTFE's strength and lubricity. The addition of Au NPs made the composites wear resistant. These films were created by coating the surfaces with the composite via the dip coating method. The concentration of Au NP was 0.07 wt%, and its thickness ranged from 0.2 to 0.5 μm. The wear resistance of the PTFE film improved twofold upon adding Au NPs at a concentration of 0.06 wt%. The durability tests were conducted under loads of 20 g and 50 g. Studies on wear progression and durability indicated that the samples had different wear mechanisms. PTFE film was often delaminated because of the poor film-to-substrate adhesion and tearing. Contrarily, PTFE+Au exhibited negligible signs of delamination. A PTFE film with Au NPs reduced the COF by up to 50% while increasing wear resistance by twofold [29]. **Ultrafine particles** are critical for improving lubricating oil performance. At 100°C, Chanaka Kumara *et al.* investigated dodecanethiol-modified Palladium (Pd) for boundary lubrication. Particle diameters ranged from 2 to 4 nm. The surface-functionalized NPs improved

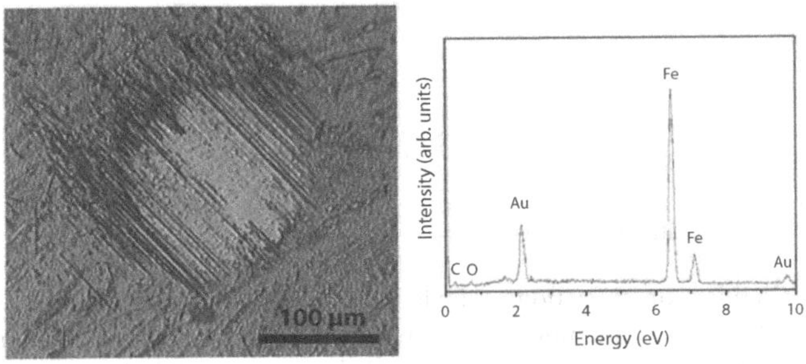

Figure 4.10 Optical micrographs showing the adhered material onto the ball surface for Au NPs and corresponding EDX analysis. Reprinted with permission from Sánchez-López *et al.* [23], Copyright (2009) @ Elsevier.

the lubrication performance by creating low-friction, anti-wear tribofilm between the contact surfaces. The modified-Pd NPs demonstrated greater long-term stable suspension in lubricating oil during tribological experiments. Interestingly, the organic ligand exhibited anti-wear functionality, but did not affect friction behavior. It was considered that the metallic core is principally responsible for friction and wear reductions. The Pd NPs successfully protected both the steel ball and the cast iron flat by considerably reducing the friction and wear. This Pd NP-induced tribofilm was dominated by Pd-S compounds, as evidenced by a combination of nanostructural inspection and chemical analyses. Furthermore, the film generated by dodecanethiol-modified palladium NPs was 10 times thicker (2–3 μm) and was responsible for the improved lubricating capabilities [62]. Figure 4.11 shows the friction coefficient traces and wear volumes for base oil

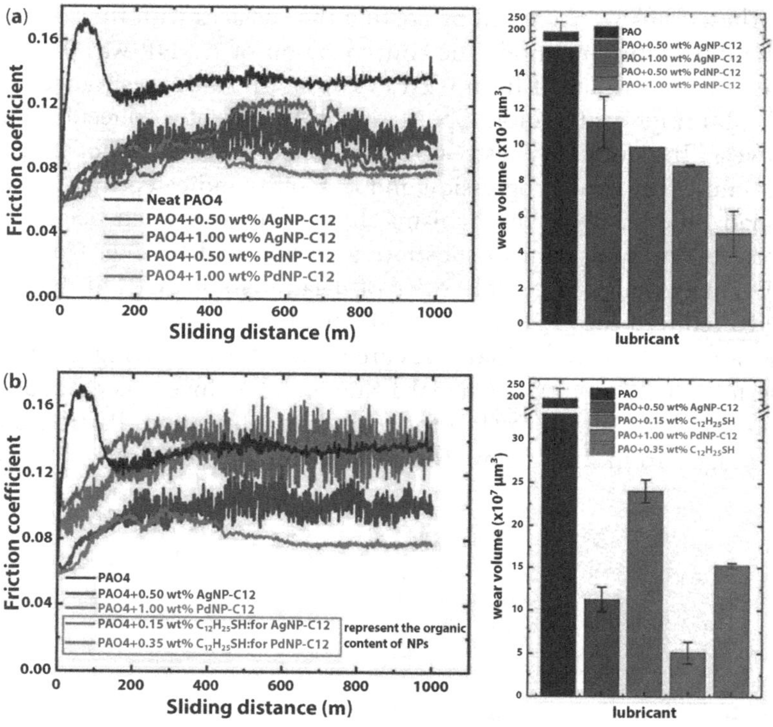

Figure 4.11 Comparison of friction coefficient traces (left) and wear volumes (right) produced by (a) the oils containing organic-modified Ag and Pd NPs at different concentrations and (b) oils containing the organic ligand alone, PdNP-C12 seems more effective than AgNP-C12 in reducing friction and wear. The organic ligand seems to contribute to the wear protection only with little impact on the friction behavior. Reprinted with permission from Kumara *et al.* [62], Copyright © 2018 ACS Publications.

and Pd-containing oil. Lubricants containing Pd NPs reduced the electrical resistance of the contact and enabled its use in tiny connections and microelectronics. Tetraalkylammonium surface-protected Pd NPs were produced and studied by L. Kolodziejczyk *et al.* [37]. The metal NPs were capped with alkylammonium molecules via van der Waals forces. It should be noted that a lower metallic content was observed in the Pd sample which indicated that the ratio of hydrocarbon coverage to metal-core was quite large. The greater carbon content in the cover layer facilitated the adsorption of surface-capped Pd NPs onto the surface of steel balls. The creation of a carbonaceous transfer layer on the steel surface occurred due to the great affinity of carbon for iron surfaces. The formulation of the lubricant was made by dispersing Pd NPs in tetrabutylammonium acetate (TBA). However, Pd produced a thick and homogenous metal-like layer that acted as a protective barrier for the surfaces and prevented direct contact between asperities. This phenomenon increased the longevity of sliding surfaces. In the case of a pure TBA solution, the resistance varies between 20 and 40 kΩ. However, the addition of Pd NPs decreased its value to about 100 Ω (a 97%–99.5% reduction). In addition, the surface-modified Pd NPs displayed a particular alignment as a result of their passage across the contact region. Interestingly, the film oriented in the sliding direction under strong contact pressure and relative velocity. This was a significant advantage compared to fluids and greases, as it made the material resistant to harsh contact situations when fluids were pushed out of the contact region.

Mineral oils offer superior rheological qualities compared to synthetic oils. R. Chou *et al.* investigated the effect of **Nickel (Ni) NPs** on the tribological properties of synthetic oil (polyalphaolefin, PAO6) at concentrations of 0.5, 1, and 2 wt%. The diameter of the NPs was 20 nm. The inclusion of Ni NPs in PAO6 decreased the COF considerably. These NPs had the superior tribological performance for the suspension containing 0.5 wt% reduced friction by 30% and wear by 54% at severe pressures. Higher Ni concentrations resulted in increased Ni deposition on the wear surface of the blocks, which consequently improved the tribological properties of the base oil [63]. **Iron (Fe)** is one of the most suitable candidates when manufacturing ease, cost of reagents, and gains in tribological characteristics are considered a priority. Tatjana Maliar *et al.* studied the tribological behavior of commercially available SAE (Society of Automotive Engineers) 10 mineral and rapeseed oil with/without surfactant/Fe particles, which were generated directly in the oil phase. By adding Fe particles at a concentration of 0.1 wt% to rapeseed oil, the COF and anti-wear properties were enhanced. The viscosity of the oil was primarily determined by the mix of oil and surfactant. The Fe particles in the range of 50 to 140 nm with oxyethylated alcohol surfactant produced the oils with the finest tribological capabilities. Mixtures of Fe and Cu NPs decreased the friction and wear of specimens far beyond pure Cu NPs, which indicated that Fe NPs were more effective in the formation of friction and wea-reducing layer [64]. Juozas Padgurskas *et al.* researched mineral oil containing Fe, Cu, and Co NPs, as well as their mixtures [21]. The combination of Fe with Co NPs reduced friction torque by 36%, while the addition of Cu reduced the friction more rapidly and to a larger extent (53%). At 2000 s of testing time, the Co+Cu and Fe+Cu combinations stabilized at a considerably lower level, and produced a tribologically active anti-friction layer, whereas the Fe+Co mixture increased the friction. Fe NPs coated with Co and Cu were figured out as the most efficient NP additive. Fe NPs coated with Cu (Fe-Cu) decreased the friction torque by 55% compared to the basic oil stabilized at 2000 s. Although, Fe-Co first demonstrated an increase in the friction and eventually dropped to 50% in comparison to the SAE 10 (control). Additionally, Fe and Co NPs reduced the wear by 23% and 11%, respectively. The combination of Cu and Fe NPs reduced the wear by 50%, and the Co+Cu and Fe+Co compositions were effective in lowering the wear by 44% and 32%, respectively. Intriguingly, the usage of Fe NPs covered with Cu and Co shells had a nearly similar effect on the endurance of the friction pair (20% less wear), although it was less effective than composites of single-metal NPs. However, the shelled NPs minimized the COF more efficiently. Tribological testing revealed Co NPs as a stand-alone were less effective compared to when combined with Fe NPs in the form of mixtures and shells. Fe NPs as a standalone additive were less effective than Cu NPs. However, in the mixture, they improved the tribological properties of Cu and Co NPs. The comparative analysis is depicted in Figure 4.12. Overall, Co NPs were excellent nanoadditives in presence of other metallic NPs.

Oxide-based NPs have unique physical and chemical properties and improve the tribological properties of oil via mechanisms depicted in Figure 4.6. Oxides of metals such as CuO, ZnO, Al_2O_3, TiO_2, and ZrO_2 have excellent lubricating qualities. They are extensively studied as nanoadditives in nanolubricants. For copper oxide $(CuO)_{two}$ NP concentrations of 1 and 2 wt% were studied and the COF was reduced by 14% and 23%, respectively. For the latter, the viscosity of nanolubricant rose by 20%. The wear volume strongly depended on the concentration of the CuO NPs as it reached a maximum of 1 wt% and a minimum of 2 wt% [65]. In an intriguing application, CuO NPs were added to coconut oil (as a base lubricant) in a modest quantity to improve its friction-reducing capabilities. The friction decreased (0.34 wt%) predominantly because of a shift from sliding to rolling friction. The flash point of oil was unaffected by the

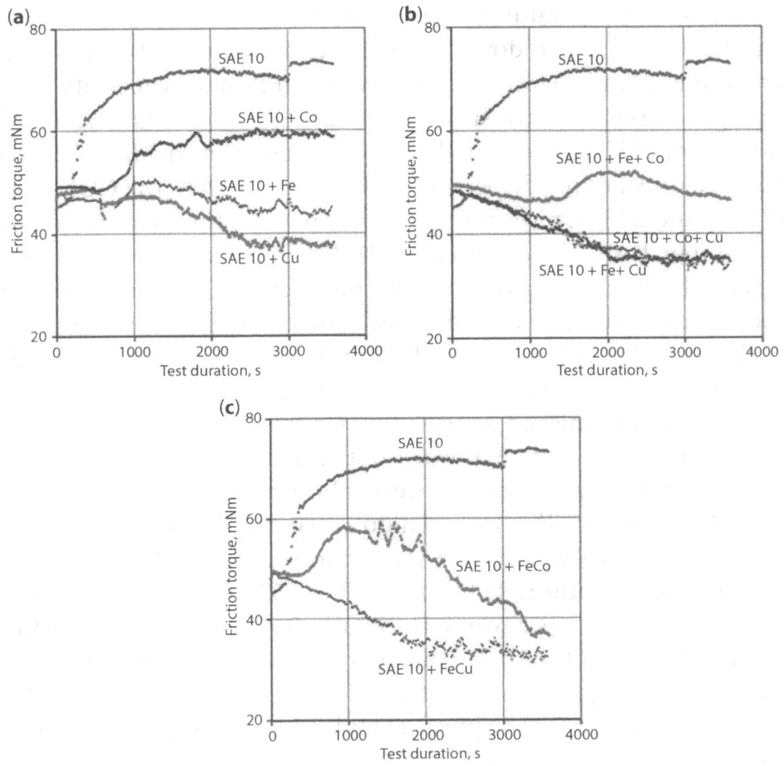

Figure 4.12 Friction torque graphs of base oil SAE 10 and oils with nano-additives: (a) with metallic NPs, (b) with mixtures of NPs, and (c) with Fe NPs coated with Cu and Co. Reprinted with permission from Padgurskas *et al.* [21], Copyright (2013) @ Elsevier.

addition of NPs. However, the fire point rose [66]. CuO NPs were the most effective in minimizing friction and anti-wear features. The reduction in friction was evaluated because of the effects of viscosity at low temperature and rolling mechanisms at high temperature. The anti-wear mechanism was related to the deposition of CuO NPs on the friction surface, which minimized the shearing stress and enhanced the tribological behavior of the base oil.

Oil-soluble ZnO NPs exhibited exceptional antioxidation, anti-wear, anti-friction, and multifunctional qualities. Its diameter was in the nanometer range. It was a good anti-wear additive that minimized metal-on-metal friction and wear [24]. When added to the control fluid (API Group III base oil + stabilizer), ZnO NPs reduced the friction by 5.2% in comparison to the control fluid. ZnO NP+oil based nanofluids successfully reduced the wear by 82.8%. However, the ZnO nanofluids were not effective at temperatures beyond 100°C because the stabilizer acted unexpectedly and broke down at higher temperatures. It also displayed shear-thinning tendencies [25]. There is a significant propensity for NPs to aggregate. Consequently, surface modification of NPs is crucial. When lubricating oil KH-560 was modified with aluminum oxide (Al_2O_3) NPs, it exhibited outstanding dispersion stability. At 0.1 wt%, the COF reduced by 17.61% and 23.9%, respectively. The diameter of the wear scar resulting from the four-ball test decreased by 41.75%. The addition of Al_2O_3 NPs enhanced the lubricating qualities by producing a self-laminating coating on the contact surfaces and transformed sliding into rolling friction [67]. Titanium Dioxide (TiO_2) NPs are yet another type of nanoadditive that demonstrated superior suspension stability. The wear of the steel ball specimens diminished as the value (in wt%) of TiO_2 was increased. The predominant wear type of worn-out surfaces was ploughing wear. At 0.075 wt%, TiO_2 acted as an anti-wear lubricant. It had a higher value of the flash temperature. This formulation enhanced the performance of the pure lubricant. At 75°C, the COF was recorded to be 0.01. It decreased by 60% at room temperature (RT) and by 80% at 75°C relative to the pure base oil. In particular, the wear scar surfaces are significantly smoother for the TiO_2-treated lubricant [68]. Zirconia (ZrO_2) possesses superior corrosion resistance, even highly concentrated acids have difficulty penetrating it. At RT, ZrO_2 offered superior physical and mechanical qualities compared to conventional advanced ceramics. ZrO_2 also possesses greater hardness, compressive strength, bending, lower fracture toughness, Young's modulus, and thermal expansion coefficient. By stabilizing ZrO_2, its lower fracture toughness can be increased. A. D. Tóth *et al.* studied ZrO_2 ceramic particles as a nanoadditive to conventional lubricants. The optimal concentration was determined to be 0.4 wt%. The lubricant sample with this NP content was able to reduce the diameter of the wear scar by 40% and the COF by 1%. At this concentration, the volume of wear was reduced by more than 86%. The most prevalent wear processes observed were abrasion and fatigue pitting. In the absence of the ZrO_2 NPs, the quantity of wear was dramatically low. The particles form agglomerates via van der Waals interactions and accumulated on the surface and within worn-out grooves, thus providing smoother contact surfaces [69].

The inclusion of oleic acid-modified copper sulfide (CuS) nanorods has substantially reduced the COF and wear rate of the pins of the pin-on-disc wear apparatus. The CuS nanorods treated with oleic acid exhibited greater anti-wear capabilities. The COF for the acid-modified nanorods dropped to half compared to the pure nanorod additives. In addition, the wear rate of the treated CuS nanorods was 10% lower than the pure nanorods.

These rods resulted in a full and uniform tribofilm formation. Overall, the addition of oleic acid-modified CuS nanorods enhanced lubricating qualities to a greater extent [70].

4.4.5 Transition Metal Dichalcogenides (TMDCs)-Based Nanolubricants

Advancements in synthesis methods and exfoliation techniques have rendered the use of monolayers of two-dimensional layered materials, such as metal dichalcogenides (TMDCs), as lubricant additives due to their unique molecular structure and lubricating properties [71–73]. These layered compounds possess structure-dependent properties that are more significant compared to their bulk counterparts, as shown in Figure 4.13. TMDCs include a large family of layered materials, which can be represented by the formula MX_2, where M is a transition metal element such as Mo, W, and X is a chalcogen atom such as S, Se, and Te. Research on these heterostructures, based on atomically thin two-dimensional TMDC crystals, offers a new approach to be used as additives to improve the friction reduction and anti-wear properties of base lubricants. Also, the advancement of exfoliating monolayers of such compounds gives us better knowledge of interaction between monolayer surfaces and materials. This also enables the researcher for homogenously dispersing monolayer in base liquid media. All these recent advancements allows the advanced research on TMDC materials to study their tribological performance as lubricant additives. The most common metal dichalcogenides, MoS_2 and WS_2, have long been recognized as excellent solid lubricants and lubricant additives [74–76]. Other metal dichalcogenides, such as $TiSe_2$ and $NbSe_2$, have been more recently reported to present satisfactory performance as additives after doping [77].

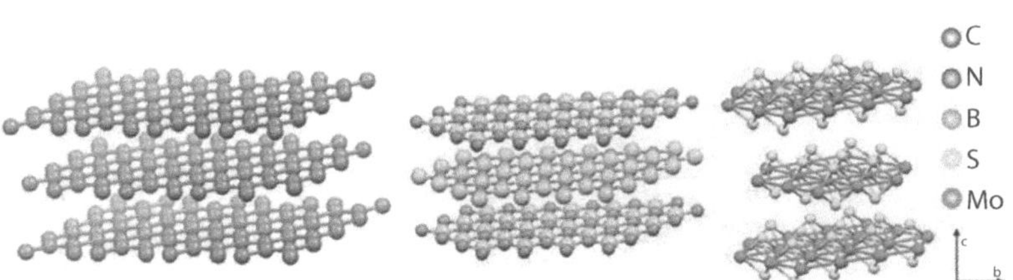

Figure 4.13 Crystal structures of archetypal layered structures: graphite and inorganic analogues of hexagonal boron nitride and molybdenum disulfide. Reprinted with permission from Tedstone *et al.* [78], Copyright (2016) @ ACS Publications.

The atomic structure of two layers of TMDC material of the form MX_2 is shown in Figure 4.14. In TMDCs, the intralayer MX bonds between the transition metal atom and the chalcogen atoms are regarded as covalent bonds, and the individual MX_2 layers are held together by relatively weaker interlayer van der Waals forces. Rhombohedral and hexagonal are the two important crystals forms for TMDCs. The hexagonal (2H) polytype are commonly used by researchers for tribological applications, although the rhombohedral (3R) polytype is also found in nature but less commonly used.

These polytypes comprise X—M—X layers stacked in alternate arrangements: parallel orientation in rhombohedral (3R) and antiparallel orientation in hexagonal (2H). The hexagonal form comprises a six-fold symmetry, two MX_2 moieties per unit cell, and a lamellar structure. The arrangement is such that each transition atom is equidistant from six chalcogen atoms, and each chalcogen atom is equidistant from three transition metal atoms. The strong bonding between transition atoms and weak bonding between layers render them useful as low friction coatings [79].

TMDCs are often used as lubricating agents either in the form of oil additives or as coatings. The coatings can be applied either as thick films fabricated by electrochemical processes [81] or as thin films using physical vapor deposition system. TMDCs could be formed as well during the sliding process on the surfaces in the contact by tribochemical reactions [82].

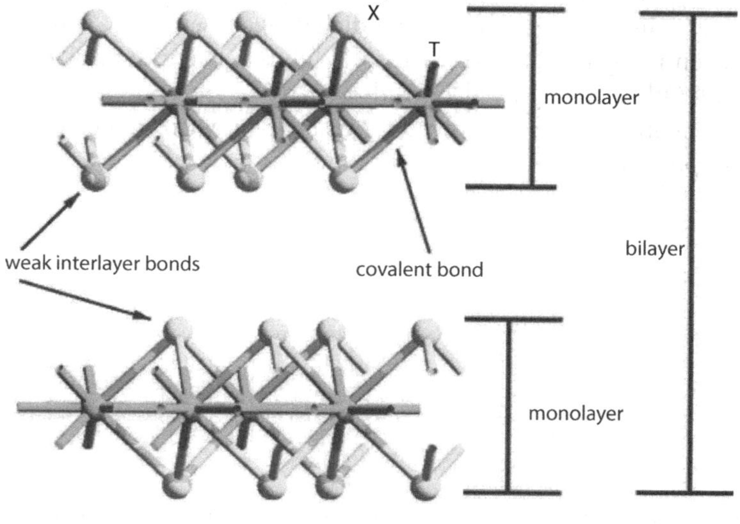

Figure 4.14 Crystal structure of TMDCs. Reprinted with permission from Kuc *et al.* [80], Copyright (2011) @ APS.

In general, TMDC heterostructures can be fabricated via either top-down methods (e.g., mechanical exfoliation) or bottom-up methods (e.g., CVD). Vertical heterostructures are formed by vertical stacking of different 2D TMDCs. The properties of such heterostructures are dominated by the stacking orientation and interlayer coupling strength. On the other hand, the lateral heterostructures can be obtained by epitaxial growth with sharp interfaces and edge contacts to offer easier band offset tuning. The anisotropic nature of TMDCs enables them to be used as exceptional solid lubricants materials having a COF lower than 0.01.

Martin *et al.* reported that the friction coefficient of MoS_2 thin films to be 0.001 under ultra-high vacuum (UHV) conditions. They proposed that the adhesive force between MoS_2 films and the underlying substrate should be higher compared to weak attractive forces between lamellae. This results in occurrence of slip between the lamellae and consequently the friction coefficient will be less [83]. Stupp *et al.* used a co-sputtering method to incorporate d-block metals such as Co, Cr, and Ni into molybdenum disulfide films. The dopants modified molybdenum disulfide films with a stable frictional response and a reduced wear rate [84]. Similarly, MoS_2 films with Ti dopants produced thin films with dense, featureless morphology and has been attributed with the improvement in tribological improvement [85]. The films were found to be amorphous, although the peaks observed by Raman spectroscopy corresponding to the E_{2g} and A_{1g} vibrational modes of 2H MoS_2 were intensified in the region of tribological wear. Vattikuti *et al.* synthesized MoS_2 nanoflowers and nanosheets and compared their lubricating effects [86]. The nanosheets exhibited superior anti-wear and friction-reducing properties as a lubrication additive in engine oil compared to the nanoflowers. The average COF and wear scar size with the nanosheet additive was significantly reduced compared to those with the nanoflower additive, altogether indicating that nanosheet structure is a better lubricant additive than nanoflower structure; the nanosheets better assisted the formation of a protective tribofilm on the rubbing surface. Weak van der Waals force and repulsive interaction due to Coulombic effect in nanosheets are responsible for the formation of a tribofilm [87]. Three-dimensional topography examination confirmed the formation and adherence of a tribofilm on the counter parts having enhance the tribological properties of the lubricant [86].

P. Nallasamy *et al.* examined the tribological properties of molybdenum disulfide (MoS_2) NPs treated with lecithin as an additive to PAO to determine its suitability for usage in machine tools. Since lecithin is a Zwitterionic surfactant, it is usually added to impede the agglomeration and settling of NPs. MoS_2 is an essential solid lubricant with superior friction and wear qualities in inert or vacuum conditions. It had been intensively explored because of its peculiar lamellar crystal structure, which promoted intrinsically lower shear strength and consequently low friction [88]. Bearing tests revealed that the formulation is a suitable candidate for high-temperature performances [89]. It should be noted that MoS_2 NPs have outstanding tribological characteristics. In grinding, the additives reduced the cutting forces. MoS_2 with canola oil and lecithin demonstrated eco-friendly properties [89]. As mentioned earlier, MoS_2 NPs have a lamellar structure and formed a tribofilm between sliding steel surfaces. These properties lead to the reduction of the COF by 24.2% at 1 wt%. For PAO + 1% MoS_2, the static and dynamic COFs were of the same order and had a lower stick-slip value when measured at 60°C, 80 °C, and 100 °C. In addition, this formulation had the maximum weld load [90]. It was observed that tungsten disulfide (WS_2) NPs resulted in a smoother and more homogenous oil formulation. The introduction of WS_2 NPs in hyper-eutectic Al-Si alloys containing carbon steel decreased the volumetric wear. This resulted in the increased viscosity of the conventional oil, which caused a reduction in wear rate to a value of 0.006313×10^{-6} N/m (0.08418×10^{-6} N/m without WS_2 NPs). Table 4.2 lists various nanoparticles added with different dispersion methods alongwith the test parameters, and test results for various types of base oils.

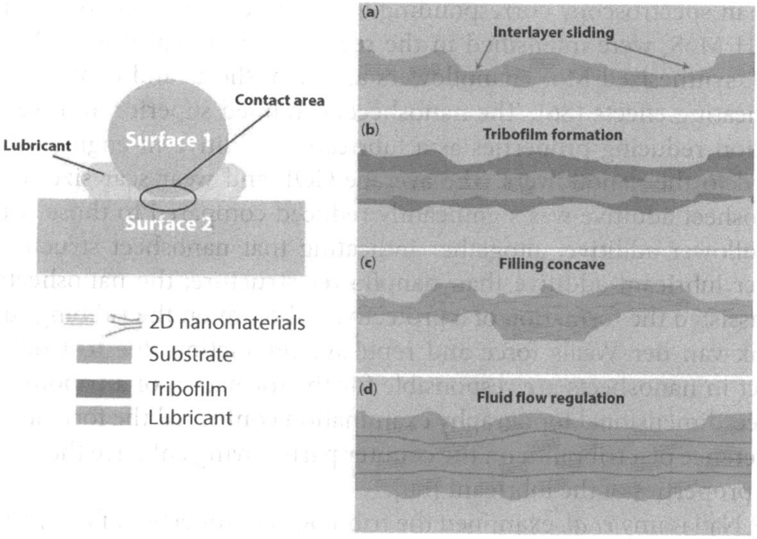

Figure 4.15 Possible mechanisms of 2D nanosheets to reduce friction and wear: (a) entering the contact area of sliding surfaces; (b) tribofilm forming; (c) filling the pits and gaps of the contact area; and (d) affecting the fluid drag and viscosity. Reprinted with permission from Xiao and Liu [92], Copyright (2017) @ Elsevier.

Table 4.2 List of nanoparticles added, dispersion method, test parameters, and test results for various types of base oils.

Details of nanolubricants						Test parameters			Test results			
NPs	Base oil	Particle size (nm)	Concentration (mg/L)	Dispersion method	Tribometer	Load	Speed	Temperature and duration	Wear reduction	Friction reduction	Optimum concentration	References
Category: Mineral Oil												
Fe	Mineral oil (SAE 10) and rapeseed oils	50-140 nm with OS/ENB surfactant	0.1 wt%	-	Four-ball wear tribotester	40 and 60 kg	1200 rpm for 1 and 2h	75°C	Fe NPs of 165-340 nm size range reduced COF more than smaller Fe NPs of 50-120 nm size	At high load, OS + Fe showed reduction of wear scar by 64%, ENB+Fe 41%	0.1 wt%	[64]
TiO_2	Mineral oil	20–25 nm	0.25, 1, 2 wt%	Mechanical stirrer, 15 min	Reciprocating pin-on-disk	14.715 N	0.05 m/s	Ambient, 30 min	Reduction of 0.01 COF		0.25 wt%	[93]
	Multi-grade engine oil SAE 20W 40	10–25 nm	1.5 wt%	Ultrasonic shaker	Pin-on-disc tribometer	40, 60, 90 N	0.5, 1.0, 1.5 m/s	5 min	Wear significantly reduced	Reduced by 50%	1.5 wt%	[94]
Al_2O_3	Pure lubricating oil	78 nm	0.05, 0.1, 0.5, 1.0 wt%	Ultrasonication, 30 min	Four ball tribometer, thrust-ring tribometer	147 N, 200 N	1450 rpm, 1200 rpm	75 °C, 30 min	Average COF reduction are 17.61% (four ball) and 23.92% (thrust-ring), WSD reduction of four ball test is 41.75%		0.1 wt%	[67]
CuO	Mineral oil	30–40 nm	0.5–1.5 wt%	Ultrasonic shaker, 30 min	Pin-on-disc tester	40 and 60 N	0.5–1.5 m/s -	-	Wear significantly reduced	Reduced up to 50% of COF	1 wt%	[95]
	Paraffin oil	50 nm	0.2, 0.25, 2 and 3 wt%	Ultrasonic bath, 1 h	Four-ball tribomachine	40 kg	1200 rpm	60–70 °C, 15 min	The higher the concentration of CuO, the better the tribological properties		-	[96]
MoS_2	SE15W40	~50 nm	0.1, 0.5, 1.0, 2.0 and 5.0 wt%	High shear homogenizer, 30 min	Disc-on-disc frictional testing machine	1500 N	500 rpm	Ambient temperature, 180 s	Tribological performance of base oil was enhanced significantly		-	[97]
Ag	Mineral oil (MO)	20 nm	0.5 , 1 and 2 wt%	Ultrasonicator	Four- ball tester	100-10000 N	300-3000 rpm	Ambient to 120°C	Average wear scar diameter: 0.601 mm, 0.587 mm, 0.557 mm, and COF: 0.36, 0.34, and 0.32 for MO+0.5% Ag, MO+1% Ag, and MO+2% Ag		MO+2% Ag	[59]
Fe, Cu, Co	SAE 10 mineral oil	50-340 nm	0.5 wt%	-	Four- ball tribotester	150 N	1420 min^{-1}	1 h	Reduced up to 11% (Co), 23% (Fe), 47% (Cu)	Reduced up to 20% (Co), 39% (Fe), 49% (Cu)	0.5 wt%	[21]

(Continued)

Table 4.2 List of nanoparticles added, dispersion method, test parameters, and test results for various types of base oils. (*Continued*)

Details of nanolubricants						Test parameters			Test results			
NPs	Base oil	Particle size (nm)	Concentration (mg/L)	Dispersion method	Tribometer	Load	Speed	Temperature and duration	Wear reduction	Friction reduction	Optimum concentration	References
ZrO_2	Neat group III base oil	-	0.1, 0.4 wt%	Ultrasonicator	Optimal SRV5 tribometer	50 N, 100 N	2h	50°C	Reduced by 40%	Reduced by 1%	0.4 wt%	[69, p. 2]
Au	Paraffin (PAR)	2 nm	5 wt %	-	Pin-on-disc tribometer	at 7 N and 15 N	10 cm/s of linear speed	Ambient	Friction (0.1), wear rate (1010 mm^3/Nm)		5 wt%	[23]
WS_2	Conventional engine oil	-	0.05 g/l	Ultrasonicator	Pin-on-disc tribometer	10, 20, 30 N	1.32, 1.885, 3 m/s, 360 min	31°C	0.00631×10^{-6} N/m compared with a wear rate of 0.08418×10-6 N/m without WS2 NPs	0.24 compared with a wear rate of 00.266 N/m without NP	0.05 g/l	[91]
Cu	Paraffin Oil	10–60 nm	0.02 wt%	Mechanical agitation and ultrasonic dispersion,30 min	Four ball machine -	300 N	1450 rpm	20 °C, 30 min	Reduction of 23%	Reduction of 26%	0.02 wt%	[57]
	Chevron Taro 30 DP 40 and Teboil Ward	80–120	3 wt%		Pin-on-disc tribometer	0.1–180 mN	0.02 mm/s	25 °C	From 0.023 to 0.018 mg	From 0.15 to 0.11	3 wt%	[56]
	SAE grade15W-40	50 nm	2.5, 5, 7.5 and 10 wt%		Ball-on-disk tribometer	50 N	10 to 30 Hz	Rt, 30 min	The wear reduced as the concentration increased until 10 wt% wear increased		7.5 wt%	[98]
Category: Synthetic Oil												
Cu	Polyalphaolfin (PAO 6)	25 nm	0.5 and 2 wt%	Ultrasonic probe, 30 min	Block and ring, four ball	medium load 165 N,	1 m/s, 1470 rpm	-	Reduction of 50% wear for 0.5 wt% and reduction of 16% wear for 2 wt%		0.5 wt%	[99]

(*Continued*)

Table 4.2 List of nanoparticles added, dispersion method, test parameters, and test results for various types of base oils. (*Continued*)

Details of nanolubricants						Test parameters			Test results			
NPs	Base oil	Particle size (nm)	Concentration (mg/L)	Dispersion method	Tribometer	Load	Speed	Temperature and duration	Wear reduction	Friction reduction	Optimum concentration	References
Ni	Polyalphaolfin (PAO 6)	20 nm	0.5, 1.0, 2.0 wt%	Ultrasonic probe, 30 min	Block on ring, four ball tester	165 N	2 m/s, 1470 rpm	-	Reduction from 7%–30%	Reduction from 5%–45%	0.5 wt%	[63]
		7.5, 13.5 and 28.5 nm	0.02, 0.05, 0.1, 0.2, 0.4 wt%	-	Four ball friction and wear tester	300 N	1450 rpm	30 min	Addition of 0.05 wt% Ni exhibited excellent anti-wear behavior		0.05 wt%	[100]
Category: Biolubricants												
Cu	Pongamia Oil	25–85 nm	0.025, 0.05,0.075 & 0.1wt%	Ultrasonicator	Pin-on-disc tribometer	40 N	200 to 800 rpm	-	Only addition of 1% nano-copper caused the further increase of wear and COF		0.075 wt%	[101]
TiO_2	Trimethylolpropane (TMP) ester	-	1 wt%	Ultrasonic bath, 8 h	Four ball wear tester	40–120 kg	1200 rpm	Rt, 10 min	Decreased by 11%	Decreased by 15%	1 wt%	[102]
	Palm oil	22.98 nm	0.05, 0.1, 0.2 wt%	Ultrasonic bath, 30 min	Four ball tribotester	40 kg	1200 rpm	60–70 °C, 15 min	Only addition of 0.1 wt% TiO of friction and wear 2 exhibited reduction		0.1 wt%	[103]
CuO	Palm kernel oil (PKO)	40 nm	0.34 wt%	High-shear homogenizer, 40 min	Pin-on-disc tribometer	9.81 N	0.2 m/s	60 min	Reduced WSD by 48%	Reduced COF by 56%	0.34 wt%	[104]

(*Continued*)

Table 4.2 List of nanoparticles added, dispersion method, test parameters, and test results for various types of base oils. (*Continued*)

Details of nanolubricants						Test parameters			Test results			
NPs	Base oil	Particle size (nm)	Concentration (mg/L)	Dispersion method	Tribometer	Load	Speed	Temperature and duration	Wear reduction	Friction reduction	Optimum concentration	References
MoS_2	Coconut oil and paraffin oil	90 nm	0.25, 0.5, 0.75, 1wt%	Ultrasonic shaker, 1 h, 50 °C	Pin-on-disc tribometer, four ball tester	2 to 4 MPA, 392 N	0.47 to 1.414 m 600 rpm /s,	75 ± 2 °C, 60 min	Friction and wear reduced with the increase of concentration until optimum concentration was reached		0.53% for coconut oil and 0.58% for paraffin oil	[42]
Category: Various tests and oils												
Cu	Lithium grease	50–100 nm	0.2, 0.5, 1.0, 2.0 wt%	Homogenized by triple-roller mill	Ball-on-disk tribometer	200 N	50 Hz	80 °C, 2 h	Reduction of 82.2% wear loss	Reduction of 12% friction	0.5 wt%	[55]
Sn and Fe	Macs base oil	30–60 nm and 20–70 nm	0.1, 0.5, 1.0 wt%	Ultrasonic probe, 5 min	Vacuum four-ball tribometer	300 N	1450 rpm	25 °C, 30 min	Sn and Fe exhibited friction and wear reduction, but Sn was effective on friction reduction and Fe was effective on anti-wear		1.0 wt%	[22]
Pd	TBA	2 nm	1–10 wt%	-	Ball-on-disc tribometer	1–20 N	10 cm/s	-	Addition of 2 wt% of Pd improved the tribological properties, above 5 wt% wear rate increase		2.0 wt%	[105]
CuO	Water- based lubricant	20 nm	0.1, 0.2, 0.4, 0.6, and 0.8 wt%	Ultrasonic dispersed, 20 min	Four-ball tribotester	147, 196, and 245 N	1440 rpm	Rt, 10 min	Addition of 0.2 wt% of CuO reduced the friction up to 69.2% and wear reduced up to 55.1% under different load		0.2, 0.4, 0.6 wt% at certain load	[106]

(*Continued*)

Table 4.2 List of nanoparticles added, dispersion method, test parameters, and test results for various types of base oils. (*Continued*)

Details of nanolubricants						Test parameters			Test results			
NPs	Base oil	Particle size (nm)	Concentration (mg/L)	Dispersion method	Tribometer	Load	Speed	Temperature and duration	Wear reduction	Friction reduction	Optimum concentration	References
TiO_2	Water-based lubricant	20 nm	0.2–8.0 wt%	Ultrasonication with stirring, 10 min	Ball-on-disk tribometer	50 N	20 mm/s	Rt, 10 min	Can be decreased by 97.8%	Can be decreased by 49.5%	0.8 wt%	[107]
	Water-based lubricant	20 nm	0.4–8.0 wt%	Ultrasonication with stirring, 10 min	Ball-on-disk tribometer	5 N	50 mm/s	25 °C, 30 min	COF of oxidized disk is lower than clean disk due to Fe element from steel ball being oxidized and then formed a protective film		4.0 wt%	[36]
	Oil-in-water	30 nm	0.3–2 wt%	Stirring and ultrasonic vibration	Ball-on-disk tribometer	50 N	50 mm/s	80°C, 30 min	Addition of 1 wt% oil COF, other concentrations were higher than base oil		1 wt% oil + 2 wt% TiO_2 reduce 17.6% 2 wt% TiO	[108]

Owing to their nanoscale features and film formation properties at low temperatures the COF of the formulation improved drastically [91].

Numerous studies have been devoted to understanding the underlying mechanisms of 2D materials as lubricant additives in effort to promote their wider application. Generally, the evolution of the protection of the lubricants can be divided into four aspects: (1) entering the contact area of sliding surfaces; (2) forming a tribofilm; (3) filling the pits and gaps of the contact area; and (4) affecting fluid drag and viscosity. The four proposed mechanisms are schematically expressed in Figure 4.15.

4.6 Conclusion

Nanolubrication is a successful strategy to lessen friction and wear. The use of NPs as lubricant additives is extremely important for energy conservation, emission reduction, and environmental protection. This chapter provides the recent advances in metal-based nanolubricants. The current study analyzed the tribological properties and lubrication mechanism of various types of NPs used as lubricant additives. These results highlighted that nanometric metallic particles are excellent materials for their use as anti-wear additives in extreme-pressure conditions. NPs can also act as nanobearing, increasing the carrying load capacity of the lubricant. The load-carrying capacity of the suspension is tested under extreme pressure conditions. Nanolubricants free from sulfur or phosphorus are eco-friendly and do not interfere with the enhancement of tribological qualities. The vast majority of published research found that NP-enriched lubricants have enhanced tribological performance. However, the lubricating process containing NPs is complex and require maintaining the homogeneity and good stability of nanolubricants for an extended period of time. In the future, numerous modification strategies must be used to explore the NP stabilization in different base oils, and to discover novel lubricant additives with the best tribological performance under harsh circumstances.

References

1. Mobarak, H.M. *et al.*, The prospects of biolubricants as alternatives in automotive applications. *Renew. Sustain. Energy Rev.*, 33, 34–43, May 2014, doi: 10.1016/j.rser.2014.01.062.
2. Jason, Y.J.J., How, H.G., Teoh, Y.H., Chuah, H.G., A study on the tribological performance of nanolubricants. *Processes*, 8, 11, 1372, 2020.

3. Hirani, H., *Fundamentals of Engineering Tribology With Applications*, Cambridge University Press, Cambridge, UK, 2016.
4. Uflyand, I.E., Zhinzhilo, V.A., Burlakova, V.E., Metal-containing nanomaterials as lubricant additives: State-of-the-art and future development. *Friction*, 7, 2, 93–116, Apr. 2019, doi: 10.1007/s40544-019-0261-y.
5. Shafi, W.K. and Charoo, M.S., NanoLubrication systems: An overview. *Mater. Today Proc.*, 5, 9, Part 3, 20621–20630, Jan. 2018, doi: 10.1016/j.matpr.2018.06.443.
6. Yu, H., Xu, Y., Shi, P., Xu, B., Wang, X., Liu, Q., Tribological properties and lubricating mechanisms of Cu nanoparticles in lubricant. *Trans. Nonferrous Met. Soc China*, 18, 3, 636–641, Jun. 2008, doi: 10.1016/S1003-6326(08)60111-9.
7. Kalin, M., Kogovšek, J., Remškar, M., Nanoparticles as novel lubricating additives in a green, physically based lubrication technology for DLC coatings. *Wear*, 303, 1, 480–485, Jun. 2013. doi: 10.1016/j.wear.2013.03.009.
8. Wu, Y.Y., Tsui, W.C., Liu, T.C., Experimental analysis of tribological properties of lubricating oils with nanoparticle additives. *Wear*, 262, 7, 819–825, Mar. 2007, doi: 10.1016/j.wear.2006.08.021.
9. Zhang, Z., Yu, L., Liu, W., Xue, Q., The effect of LaF3 nanocluster modified with succinimide on the lubricating performance of liquid paraffin for steelon-steel system. *Tribol. Int.*, 34, 2, 83–88, Feb. 2001, doi: 10.1016/S0301-679X(00)00135-3.
10. Yadav, T.P., Yadav, R.M., Singh, D.P., Mechanical milling: A top down approach for the synthesis of nanomaterials and nanocomposites. *Nanosci. Nanotechnol.*, 2, 3, 22–48, 2012.
11. Li, D. and Xia, Y., Electrospinning of nanofibers: Reinventing the wheel? *Adv. Mater.*, 16, 14, 1151–1170, 2004, doi: 10.1002/ adma.200400719.
12. Kuo, C.-W., Shiu, J.-Y., Cho, Y.-H., Chen, P., Fabrication of large-area periodic nanopillar arrays for nanoimprint lithography using polymer colloid masks. *Adv. Mater.*, 15, 13, 1065–1068, 2003, doi: 10.1002/adma.200304824.
13. Ohring, M., *The Materials Science of Thin Films*, Academic Press, San Diego, California, 2013, doi: 10.1016/ C2009-0-22199-4.
14. Feng, S. and Xu, R., New materials in hydrothermal synthesis. *Acc. Chem. Res.*, 34, 3, 239–247, Mar. 2001, doi: 10.1021/ar0000105.
15. Gummow, R.J. and He, Y., Recent progress in the development of Li2MnSiO4 cathode materials. *J. Power Sources*, 253, 315–331, May 2014. doi: 10.1016/j.jpowsour.2013.11.082.
16. Kumar, A., Kumar, N., Sharma, Y., One-pot synthesis of pure phase Mn3O4 at room temperature and probing its long-term supercapacitive performance. *Ionics*, 25, 2, 707–718, Feb. 2019, doi: 10.1007/s11581-018-2615-2.

17. Behrens, M., Coprecipitation: An excellent tool for the synthesis of supported metal catalysts–from the understanding of the well known recipes to new materials. *Catal. Today*, 246, 46–54, May 2015. doi: 10.1016/j.cattod.2014.07.050.
18. Kolthoff, I.M., Theory of coprecipitation. the formation and properties of crystalline precipitates. *J. Phys. Chem.*, 36, 3, 860–881, Mar. 1932, doi: 10.1021/j150333a008.
19. Zhao, J., Huang, Y., He, Y., Shi, Y., Nanolubricant additives: A review. *Friction*, 9, 5, 891–917, Oct. 2021, doi: 10.1007/s40544-020-0450-8.
20. Hisakado, T., Tsukizoe, T., Yoshikawa, H., Lubrication mechanism of solid lubricants in oils. *J. Lubr. Technol.*, 105, 2, 245–252, Apr. 1983. doi: 10.1115/1.3254585.
21. Padgurskas, J., Rukuiza, R., Prosyčevas, I., Kreivaitis, R., Tribological properties of lubricant additives of Fe, Cu and Co nanoparticles. *Tribol. Int.*, 60, 224–232, 2013, doi: 10.1016/J.TRIBOINT.2012.10.024.
22. Zhang, S., Hu, L., Feng, D., Wang, H., Anti-wear and friction-reduction mechanism of Sn and Fe nanoparticles as additives of multialkylated cyclopentanes under vacuum condition. *Vacuum*, 87, 75–80, Jan. 2013, doi: 10.1016/J.VACUUM.2012.07.009.
23. Sánchez-López, J.C., Abad, M.D., Kolodziejczyk, L., Guerrero, E., Fernández, A., Surface-modified Pd and Au nanoparticles for anti-wear applications. *Tribol. Int.*, 44, 6, 720–726, Jun. 2011. doi: 10.1016/j.triboint.2009.12.013.
24. Xue, W.G., Zhao, Z.H., Wang, P., Jin, Z.L., Xu, X.H., Zhou, X.G., Performance study of zinc oxide nanoparticles for lubricant oil. *Adv. Mater. Res.*, 1118, 195–204, Jul. 2015. doi: 10.4028/www.scientific.net/AMR.1118.195.
25. Harta, I.C. *et al.*, Tribological performance of ZnO-oil nanofluids at elevated temperatures. *SAE Int. J. Fuels Lubr.*, 6, 1, 126–131, Apr. 2013, doi: 10.4271/2013-01-1219.
26. Ma, S., Zheng, S., Cao, D., Guo, H., Anti-wear and friction performance of ZrO2 nanoparticles as lubricant additive. *Particuology*, 8, 5, 468–472, Oct. 2010, doi: 10.1016/j.partic.2009.06.007.
27. An, V., Irtegov, Y., de Izarra, C., Study of tribological properties of nanolamellar WS2 and MoS2 as additives to lubricants. *J. Nanomater.*, 2014, e865839, Oct. 2014, doi: 10.1155/2014/865839.
28. Wang, H., Liu, Y., Chen, Z., Wu, B., Xu, S., Luo, J., Layered double hydroxide nanoplatelets with excellent tribological properties under high contact pressure as water-based lubricant additives. *Sci. Rep.*, 6, 1, 1–24, 2016, doi: 10.1038/srep22748.
29. Beckford, S., Cai, J., Chen, J., Zou, M., Use of Au nanoparticle-filled ptfe films to produce low-friction and low-wear surface coatings. *Tribol. Lett.*, 56, 2, 223–230, Nov. 2014, doi: 10.1007/s11249-014-0402-4.

30. Kotia, A., Haldar, A., Kumar, R., Deval, P., Ghosh, S.K., Effect of copper oxide nanoparticles on thermophysical properties of hydraulic oil-based nanolubricants. *J. Braz. Soc Mech. Sci. Eng.*, 39, 1, 259–266, Jan. 2017. doi: 10.1007/s40430-016-0664-x.
31. Gulzar, M. *et al.*, Tribological performance of nanoparticles as lubricating oil additives. *J. Nanopart. Res.*, 18, 8, 223, Aug. 2016, doi: 10.1007/s11051-016-3537-4.
32. Hu, Z.S. *et al.*, Preparation and tribological properties of nanometer magnesium borate as lubricating oil additive. *Wear*, 252, 5, 370–374, Mar. 2002, doi: 10.1016/S0043-1648(01)00862-6.
33. Liu, G., Li, X., Qin, B., Xing, D., Guo, Y., Fan, R., Investigation of the mending effect and mechanism of copper nano-particles on a tribologically stressed surface. *Tribol. Lett.*, 17, 4, 961–966, Nov. 2004, doi: 10.1007/s11249-004-8109-6.
34. Tang, Z. and Li, S., A review of recent developments of friction modifiers for liquid lubricants (2007–present). *Curr. Opin. Solid State Mater. Sci.*, 18, 3, 119–139, Jun. 2014, doi: 10.1016/j.cossms.2014.02.002.
35. Peng, D.X., Kang, Y., Hwang, R.M., Shyr, S.S., Chang, Y.P., Tribological properties of diamond and SiO2 nanoparticles added in paraffin. *Tribol. Int.*, 42, 6, 911–917, Jun. 2009, doi: 10.1016/j.triboint.2008.12.015.
36. Wu, H. *et al.*, Friction and wear characteristics of TiO2 nano-additive water-based lubricant on ferritic stainless steel. *Tribol. Int.*, 117, 24–38, Jan. 2018, doi: 10.1016/J.TRIBOINT.2017.08.011.
37. Kolodziejczyk, L., Martínez-Martínez, D., Rojas, T.C., Fernández, A., Sánchez-López, J.C., Surface-modified Pd nanoparticles as a superior additive for lubrication. *J. Nanopart. Res.*, 9, 4, 639–645, Aug. 2007, doi: 10.1007/s11051-006-9124-3.
38. Ji, X., Chen, Y., Zhao, G., Wang, X., Liu, W., Tribological properties of CaCO3 nanoparticles as an additive in lithium grease. *Tribol. Lett.*, 41, 1, 113–119, Jan. 2011, doi: 10.1007/s11249-010-9688-z.
39. Singh, A.P., Dwivedi, R.K., Suhane, A., Influence of nano particles on the performance parameters of lube oil–a review. *Mater. Res. Express*, 8, 10, 102001, Oct. 2021, doi: 10.1088/2053-1591/ac2add.
40. Kato, H. and Komai, K., Tribofilm formation and mild wear by tribo-sintering of nanometer-sized oxide particles on rubbing steel surfaces. *Wear*, 262, 1, 36–41, Jan. 2007, doi: 10.1016/j.wear.2006.03.046.
41. Song, X., Zheng, S., Zhang, J., Li, W., Chen, Q., Cao, B., Synthesis of monodispersed ZnAl2O4 nanoparticles and their tribology properties as lubricant additives. *Mater. Res. Bull.*, 47, 12, 4305–4310, Dec. 2012. doi: 10.1016/j.materresbull.2012.09.013.

42. Koshy, C.P., Rajendrakumar, P.K., Thottackkad, M.V., Evaluation of the tribological and thermo-physical properties of coconut oil added with MoS2 nanoparticles at elevated temperatures. *Wear*, 330–331, 288–308, May 2015, doi: 10.1016/J.WEAR.2014.12.044.
43. Kumar Dubey, M., Bijwe, J., Ramakumar, S.S.V., PTFE based nano-lubricants. *Wear*, 306, 1, 80–88, Aug. 2013. doi: 10.1016/j.wear.2013.06.020.
44. Lee, G.-J. and Rhee, C.K., Enhanced thermal conductivity of nanofluids containing graphene nanoplatelets prepared by ultrasound irradiation. *J. Mater. Sci.*, 49, 4, 1506–1511, Feb. 2014, doi: 10.1007/s10853-013-7831-6.
45. Mosleh, M., Atnafu, N.D., Belk, J.H., Nobles, O.M., Modification of sheet metal forming fluids with dispersed nanoparticles for improved lubrication. *Wear*, 267, 5, 1220–1225, Jun. 2009. doi: 10.1016/j.wear.2008.12.074.
46. Demas, N.G., Timofeeva, E.V., Routbort, J.L., Fenske, G.R., Tribological effects of BN and MoS2 nanoparticles added to polyalphaolefin oil in piston skirt/cylinder liner tests. *Tribol. Lett.*, 47, 1, 91–102, Jul. 2012, doi: 10.1007/s11249-012-9965-0.
47. Thottackkad, M.V., Rajendrakumar, P.K., Prabhakaran, N.K., Tribological analysis of surfactant modified nanolubricants containing CeO2 nanoparticles. *Tribol.-Mater. Surf. Interfaces*, 8, 3, 125–130, Sep. 2014. doi: 10.1179/1751584X13Y.0000000051.
48. Agarwal, D.K., Vaidyanathan, A., Sunil Kumar, S., Synthesis and characterization of kerosene–alumina nanofluids. *Appl. Therm. Eng.*, 60, 1, 275–284, Oct. 2013, doi: 10.1016/j.applthermaleng.2013.06.049.
49. Rivera-Solorio, C., II, Payán-Rodríguez, L.A., García-Cuéllar, A.J., Ramón-Raygoza, E.D., Cadena-de-la-Peña, N.L., Medina-Carreón, D., Formulation techniques for nanofluids. *Recent Pat. Nanotechnol.*, 7, 3, 208–215, Nov. 2013.
50. Jendrzej, S., Gökce, B., Barcikowski, S., Colloidal stability of metal nanoparticles in engine oil under thermal and mechanical load. *Chem. Eng. Technol.*, 40, 9, 1569–1576, 2017, doi: 10.1002/ceat.201600541.
51. Li, D., He, Q., Li, J., Smart core/shell nanocomposites: Intelligent polymers modified gold nanoparticles. *Adv. Colloid Interface Sci.*, 149, 1–2, 28–38, Jul. 2009, doi: 10.1016/j.cis.2008.12.007.
52. Zhang, L., Chen, L., Wan, H., Chen, J., Zhou, H., Synthesis and tribological properties of stearic acid-modified anatase (TiO2) nanoparticles. *Tribol. Lett.*, 41, 2, 409–416, Feb. 2011, doi: 10.1007/s11249-010-9724-z.
53. Corporation, N., *Base Oil Groups Explained*, Noria Corp., Tulsa, USA, 2012. [Online]. Available: https://www.machinerylubrication.com/Read/29113/base-oil-groups.
54. Gondolini, A., Mercadelli, E., Zin, V., Barison, S., Sanson, A., Easy preparation method of stable copper-based nanoparticle suspensions in lubricant engine oil. *Lubr. Sci.*, 32, 5, 205–217, Aug. 2020, doi: 10.1002/ ls.1496.
55. Wang, J., Guo, X., He, Y., Jiang, M., Sun, R., The synthesis and tribological characteristics of triangular copper nanoplates as a grease additive. *RSC Adv.*, 7, 64, 40249–40254, Aug. 2017, doi: 10.1039/C7RA07252J.

56. Scherge, M., Böttcher, R., Kürten, D., Linsler, D., Multi-phase friction and wear reduction by copper nanoparticles. *Lubricants*, 4, 4, 36, Oct. 2016, doi: 10.3390/LUBRICANTS4040036.
57. Yang, G., Zhang, Z., Zhang, S., Yu, L., Zhang, P., Hou, Y., Preparation and characterization of copper nanoparticles surface-capped by alkanethiols. *Surf. Interface Anal.*, 45, 11–12, 1695–1701, Nov. 2013, doi: 10.1002/SIA.5309.
58. Kumara, C., Luo, H., Leonard, D.N., Meyer, H.M., Qu, J., Organic-modified silver nanoparticles as lubricant additives. *ACS Appl. Mater. Interfaces*, 9, 42, 37227–37237, Oct. 2017, doi: 10.1021/acsami.7b13683.
59. Prabu, L., Saravanakumar, N., Rajaram, G., Influence of Ag nanoparticles for the anti-wear and extreme pressure properties of the mineral oil based nano-cutting fluid. *Tribol. Ind.*, 40, 3, 440–447, Sep. 2018, doi: 10.24874/ti.2018.40.03.10.
60. Khan, M.S., Sisodia, M.S., Gupta, S., Feroskhan, M., Kannan, S., Krishnasamy, K., Measurement of tribological properties of Cu and Ag blended coconut oil nanofluids for metal cutting. *Eng. Sci. Technol. Int. J.*, 22, 6, 1187–1192, Dec. 2019, doi: 10.1016/j.jestch.2019.04.005.
61. Maharaj, D. and Bhushan, B., Effect of spherical Au nanoparticles on nanofriction and wear reduction in dry and liquid environments. *Beilstein J. Nanotechnol.*, 3, 759–772, Nov. 2012, doi: 10.3762/bjnano.3.85.
62. Kumara, C., Leonard, D.N., Meyer, H.M., Luo, H., Armstrong, B.L., Qu, J., Palladium nanoparticle-enabled ultrathick tribofilm with unique composition. *ACS Appl. Mater. Interfaces*, 10, 37, 31804–31812, Sep. 2018, doi: 10.1021/acsami.8b11213.
63. Chou, R., Battez, A.H., Cabello, J.J., Viesca, J.L., Sagastume, A., Tribological behavior of polyalphaolefin with the addition of nickel nanoparticles. *Tribol. Int.*, 43, 12, 2327–2332, Dec. 2010, doi: 10.1016/J.TRIBOINT.2010.08.006.
64. Maliar, T., Achanta, S., Cesiulis, H., Drees, D., Tribological behaviour of mineral and rapeseed oils containing iron particles. *Ind. Lubr. Tribol.*, 67, 4, 308–314, Jun. 2015, doi: 10.1108/ILT-05-2013-0058.
65. Ghaednia, H., Jackson, R.L., Khodadadi, J.M., Experimental analysis of stable CuO nanoparticle enhanced lubricants. *J. Exp. Nanosci.*, 10, 1, 1–18, Jan. 2015, doi: 10.1080/17458080.2013.778424.
66. Thottackkad, M.V., Perikinalil, R.K., Kumarapillai, P.N., Experimental evaluation on the tribological properties of coconut oil by the addition of CuO nanoparticles. *Int. J. Precis. Eng. Manuf.*, 13, 1, 111–116, Jan. 2012, doi: 10.1007/s12541-012-0015-5.
67. Luo, T., Wei, X., Huang, X., Huang, L., Yang, F., Tribological properties of Al2O3 nanoparticles as lubricating oil additives. *Ceram. Int.*, 40, 5, 7143–7149, Jun. 2014, doi: 10.1016/J.CERAMINT.2013.12.050.

68. Birleanu, C., Pustan, M., Cioaza, M., Molea, A., Popa, F., Contiu, G., Effect of TiO2 nanoparticles on the tribological properties of lubricating oil: An experimental investigation. *Sci. Rep.*, 12, 1, 5201, Dec. 2022. doi: 10.1038/s41598-022-09245-2.
69. Tóth, Á., Szabó, Á.I., Kuti, R., Tribological properties of nano-sized ZrO2 ceramic particles in automotive lubricants. *FME Trans.*, 49, 1, 36–43, 2021, doi: 10.5937/fme2101036T.
70. Chen, L. and Zhu, D., Preparation and tribological properties of unmodified and oleic acid-modified CuS nanorods as lubricating oil additives. *Ceram. Int.*, 43, 5, 4246–4251, Apr. 2017. doi: 10.1016/j.ceramint.2016.12.066.
71. Pimentel, J.V., Polcar, T., Cavaleiro, A., "Structural, mechanical and tribological properties of Mo–S–C solid lubricant coating. *Surf. Coat. Technol.*, 205, 10, 3274–3279, Feb. 2011, doi: 10.1016/j.surfcoat.2010.11.043.
72. Polcar, T. and Cavaleiro, A., Self-adaptive low friction coatings based on transition metal dichalcogenides. *Thin Solid Films*, 519, 12, 4037–4044, Apr. 2011, doi: 10.1016/j.tsf.2011.01.180.
73. Polcar, T. and Cavaleiro, A., Review on self-lubricant transition metal dichalcogenide nanocomposite coatings alloyed with carbon. *Surf. Coat. Technol.*, 206, 4, 686–695, Nov. 2011, doi: 10.1016/j.surfcoat.2011.03.004.
74. Maritsa, L., Martel, S., Barros, R., Bol, A., Aparicio, S., Additivation of MoS2 nanosheets to synthetic poly-alpha-olefins base oils: A theoretical study of nanolubrication. *J. Mol. Liq.*, 332, 115881, Jun. 2021, doi: 10.1016/j.molliq.2021.115881.
75. Sade, H., Moshkovich, A., Lellouche, J.-P., Rapoport, L., Testing of WS2 nanoparticles functionalized by a humin-like shell as lubricant additives. *Lubricants*, 6, 1, 1–14, 2018. doi: 10.3390/lubricants6010003.
76. Stefanov, M., Enyashin, A.N., Heine, T., Seifert, G., Nanolubrication: How do MoS2-based nanostructures lubricate? *J. Phys. Chem. C*, 112, 46, 17764–17767, Nov. 2008, doi: 10.1021/jp808204n.
77. Shi, Q. *et al.*, Synthesis and tribological properties of TiSe2 nanoparticles. *J. Nano Res.*, 45, 34–41, 2017, doi: 10.4028/www.scientific.net/JNanoR.45.34.
78. Tedstone, A.A., Lewis, D.J., O'Brien, P., Synthesis, properties, and applications of transition metal-doped layered transition metal dichalcogenides. *Chem. Mater.*, 28, 7, 1965–1974, Apr. 2016, doi: 10.1021/acs.chemmater.6b00430.
79. Irving, B.J., Nicolini, P., Polcar, T., On the lubricity of transition metal dichalcogenides: An ab initio study. *Nanoscale*, 9, 5597–5607, 17, May 2017, doi: 10.1039/C7NR00925A.
80. Kuc, A., Zibouche, N., Heine, T., Influence of quantum confinement on the electronic structure of the transition metal sulfide TS${}_{2}$. *Phys. Rev. B*, 83, 24, 245213, Jun. 2011, doi: 10.1103/PhysRevB.83.245213.
81. Cardinal, M.F., Castro, P.A., Baxi, J., Liang, H., Williams, F.J., Characterization and frictional behavior of nanostructured Ni–W–MoS2

composite coatings. *Surf. Coat. Technol.*, 204, 1, 85–90, Sep. 2009, doi: 10.1016/j.surfcoat.2009.06.037.
82. Joly-Pottuz, L., Dassenoy, F., Belin, M., Vacher, B., Martin, J.M., Fleischer, N., Ultralow-friction and wear properties of IF-WS2 under boundary lubrication. *Tribol. Lett.*, 18, 4, 477–485, Apr. 2005, doi: 10.1007/ s11249-005-3607-8.
83. Martin, J.M., Donnet, C., Le Mogne, Th., Epicier, Th., Superlubricity of molybdenum disulphide. *Phys. Rev. B*, 48, 14, 10583–10586, Oct. 1993, doi: 10.1103/PhysRevB.48.10583.
84. Stupp, B.C., Synergistic effects of metals co-sputtered with MoS2. *Thin Solid Films*, 84, 3, 257–266, Oct. 1981, doi: 10.1016/0040-6090(81)90023-7.
85. Arslan, E., Bülbül, F., Alsaran, A., Celik, A., Efeoglu, I., The effect of deposition parameters and Ti content on structural and wear properties of MoS2Ti coatings. *Wear*, 259, 7 814–819, Jul. 2005. doi: 10.1016/j.wear.2005.03.004.
86. Vattikuti, S.V.P. and Byon, C., Synthesis and characterization of molybdenum disulfide nanoflowers and nanosheets: Nanotribology. *J. Nanomater.*, 2015, e710462, Oct. 2015, doi: 10.1155/2015/710462.
87. Kalin, M., Kogovšek, J., Remškar, M., Mechanisms and improvements in the friction and wear behavior using MoS2 nanotubes as potential oil additives. *Wear*, 280–281, 36–45, Mar. 2012. doi: 10.1016/j. wear.2012.01.011.
88. Bai, Y. *et al.*, Effect of MoS2 on the tribological properties of in situ Fe3Al20 wt% Al2O3-4 wt% Cr composites. *Proc. Inst. Mech. Eng. J. J. Eng. Tribol.*, vol, 1018–1029, Sep. 2013, doi: 10.1177/1350650113479249.
89. Hampson, M.R., Roberts, E.W., Cropper, M.D., Watters, R.B., Forster, D.J., Towards the effective solid lubrication of ball bearings operating at high temperature. *Proc. Inst. Mech. Eng. J. J. Eng. Tribol.*, 222, 8, 1041–1049, Aug. 2008, doi: 10.1243/13506501JET418.
90. Nallasamy, P., Saravanakumar, N., Nagendran, S., Suriya, E., Yashwant, D., Tribological investigations on MoS_2-based nanolubricant for machine tool slideways. *Proc. Inst. Mech. Eng. J. J. Eng. Tribol.*, 229, 5, 559–567, May 2015, doi: 10.1177/1350650114556394.
91. Al-Samarai, R.A., Al-Douri, Y., Ahmad, K.R., Tribological properties of WS2 nanoparticles lubricants on aluminum-silicon alloy and carbon steels. *Walailak J.*, 10, 3, 277–287, 2013.
92. Xiao, H. and Liu, S., 2D nanomaterials as lubricant additive: A review. *Mater. Des.*, 135, 319–332, Dec. 2017. doi: 10.1016/j.matdes.2017.09.029.
93. Ingole, S., Charanpahari, A., Kakade, A., Umare, S.S., Bhatt, D.V., Menghani, J., Tribological behavior of nano TiO2 as an additive in base oil. *Wear*, 301, 1, 776–785, Apr. 2013, doi: 10.1016/j.wear.2013.01.037.
94. Laad, M., Ponnamma, D., Sadasivuni, K.K., Tribological studies of nanomodified mineral based multi-grade engine oil. *Int. J. Appl. Eng. Res.*, 12, 2855–2861, 2017.

95. Jatti, V.S. and Singh, T.P., Copper oxide nano-particles as friction-reduction and anti-wear additives in lubricating oil. *J. Mech. Sci. Technol.*, 29, 2, 793–798, Feb. 2015, doi: 10.1007/S12206-015-0141-Y.
96. Asrul, M., Zulkifli, N.W.M., Masjuki, H.H., Kalam, M.A., Tribological properties and lubricant mechanism of nanoparticle in engine oil. *Proc. Eng.*, 68, 320–325, Jan. 2013, doi: 10.1016/J.PROENG.2013.12.186.
97. Wan, Q., Jin, Y., Sun, P., Ding, Y., Rheological and tribological behaviour of lubricating oils containing platelet MoS2 nanoparticles. *J. Nanopart. Res.*, 16, 5, 1–9, Apr. 2014, doi: 10.1007/S11051-014-2386-2/ FIGURES/6.
98. Sen Zhang, B., Xu, B.S., Xu, Y., Gao, F., Shi, P.J., Wu, Y.X., CU nanoparticles effect on the tribological properties of hydrosilicate powders as lubricant additive for steel–steel contacts. *Tribol. Int.*, 44, 7–8, 878–886, Jul. 2011, doi: 10.1016/J.TRIBOINT.2011.03.002.
99. Viesca, J.L., Hernández Battez, A., González, R., Chou, R., Cabello, J.J., Antiwear properties of carbon-coated copper nanoparticles used as an additive to a polyalphaolefin. *Tribol. Int.*, 44, 7–8, 829–833, Jul. 2011, doi: 10.1016/J.TRIBOINT.2011.02.006.
100. Chen, Y., Zhang, Y., Zhang, S., Yu, L., Zhang, P., Zhang, Z., Preparation of nickel-based nanolubricants via a facile in situ one-step route and investigation of their tribological properties. *Tribol. Lett.*, 51, 1, 73–83, Jul. 2013, doi: 10.1007/S11249-013-0148-4/TABLES/2.
101. Rajubhai, V.H., Singh, Y., Suthar, K., Surana, A.R., Friction and wear behavior of Al-7% Si alloy pin under pongamia oil with copper nanoparticles as additives. *Mater. Today Proc.*, 25, 695–698, Jan. 2020, doi: 10.1016/J.MATPR.2019.08.149.
102. Zulkifli, N.W.M., Kalam, M.A., Masjuki, H.H., Yunus, R., Experimental analysis of tribological properties of biolubricant with nanoparticle additive. *Proc. Eng.*, 68, 152–157, Jan. 2013. doi: 10.1016/J.PROENG.2013.12.161.
103. Shaari, M.Z., Nik Roselina, N.R., Kasolang, S., Hyie, K.M., Murad, M.C., Bakar, M.A.A., "Investigation of tribological properties of palm oil biolubricant modified nanoparticles. *J. Teknol.*, 76, 9, 69–73, 2015. doi: 10.11113/JT.V76.5654.
104. Azman, N.F., Samion, S., Sot, M.N.H.M., Investigation of tribological properties of CuO/palm oil nanolubricant using pin-on-disc tribotester. *Green Mater.*, 6, 1, 30–37, Mar. 2018, doi: 10.1680/ JGRMA.17.00026/ASSET/IMAGES/SMALL/JGRMA6-0030-F6.GIF.
105. Abad, M.D. and Sánchez-López, J.C., Tribological properties of surface-modified Pd nanoparticles for electrical contacts. *Wear*, 297, 1–2, 943–951, Jan. 2013, doi: 10.1016/J.WEAR.2012.11.009.
106. Yang, P., Zhao, X., Liu, Y., Lai, X., Preparation and tribological properties of dual-coated CuO nanoparticles as water based lubricant additives. *J. Nanosci. Nanotechnol.*, 16, 9, 9683–9689, Sep. 2016, doi: 10.1166/ JNN.2016.12394.

107. Wu, H. *et al.*, A study of the tribological behaviour of TiO2 nano-additive water-based lubricants. *Tribol. Int.*, 109, 398–408, May 2017, doi: 10.1016/J.TRIBOINT.2017.01.013.
108. Xia, W. *et al.*, Analysis of oil-in-water based nanolubricants with varying mass fractions of oil and TiO2 nanoparticles. *Wear*, 396–397, 162–171, Feb. 2018, doi: 10.1016/J.WEAR.2017.02.031.

5

Transition Metal-Based Catalysts for Preparing Biomass-Based Lubricating Oils

Binitendra Naath Mongal[1]* and Himanshu Arora[2]†

[1]*School of Engineering and Sciences, GD Goenka University, Sohna Road, Haryana, India*
[2]*Department of Chemistry, University of Allahabad, Prayagraj, Uttar Pradesh, India*

Abstract
Transition metal complexes are well known for their catalytic role in carrying out various organic transformations. As these catalysts exhibit better selectivity, stereospecificity and regiospecificity, they find application in transformations of biomass into biolubricants. The biomass-derived lubricating oils are in huge demand due to their biodegradable nature, in addition to their non-toxicity and metal-adhering properties. Further introduction of nanoparticles in the synthesized biomass-derived oil can lead to better reduction of friction. The bottleneck here in further development is the aggregation of nanoparticles in the base oil. The present chapter comprises of a systematic review of the scope of both homogeneous transition metal-based catalysts and heterogeneous systems with hybrid materials towards catalytic esterification, transesterification, hydrogenation, and hydrogenation–esterification reactions for the preparation of biolubricants.

***Keywords*:** Biolubricants, stereospecificity, friction, biomass

5.1 Introduction

The role of lubricants is to reduce the friction between two layers in contact [1]. Their mode of action is to form an intermittent layer of the lubricating material between the two surfaces in contact. The lubricant can form a

**Corresponding author*: binitendra@gmail.com
†*Corresponding author*: himanshuiitk2004@gmail.com

Mohd Yusuf, Lalit Prasad and Shafat Ahmad Khan (eds.) Nanolubricants: Generation and Applications, (135–148) © 2024 Scrivener Publishing LLC

thin, slippery layer between the sliding contact surfaces, thereby smoothing out the surface aberrations and consequent reduction of friction [2]. Lubrication enhances the life cycle of tools and machinery by reducing surface deformation due to wear and tear. Moreover, proper lubrication also prevents corrosion in machine parts. Petroleum-based oils are conventionally used as lubricants in automobiles and various industries. The constant depletion of mineral oils has prompted scientists to consider the use of sustainable biomass-based lubricants [3]. These oils when exposed to the environment cause pollution of ground water, soil, and air [4]. Furthermore, the burning of mineral oil releases harmful greenhouse gases like carbon monoxide, carbon dioxide, sulfur oxides, and nitrous oxides along with trace metals in the atmosphere [5]. Mineral oil has been reported to be carcinogenic, and continuous exposure damages the human respiratory system [6, 7]. On the contrary, ester-based biolubricants are more environmentally friendly, renewable, and non-toxic without any greenhouse emissions [8]. These lubricants are suitable for thick film, thin film, as well as high-pressure lubrication. Ester-based biolubricants are usually synthesized by the esterification reaction of fatty acids with long carbon chains and polyols' with catalytic amounts of inorganic acid or chemical compounds behaving as Lewis acids. The reaction proceeds via transesterification reaction in the presence of an acid or a base [9, 10]. Ester-based biolubricants can be further improved for high stress usage and enhanced stability by hydrogenation–esterification, the reaction undertaken with catalysts in a hydrogenating environment at high temperature and pressure. This method removes the unsaturation present in the bio-ester chain leading to saturated products with long term stability. Such methods pave the way for biolubricants to become a viable alternative to mineral oil-based lubricants [11, 12].

To our knowledge, however, there has not been a systematic report highlighting the utilization of transition metal complexes as homogeneous catalysts in biolubricant synthesis [13, 14]. Transition metal complex-based homogeneous catalysts are effective catalysts for manufacturing esters, facilitating the ease of synthesis, and high molecular dispersion [15]. However, in the case of long-chain/polyhydric alcohols, homogeneous catalysts are seldom used due to the formation of undesired by-products like soap in the process [16]. Therefore, this work provides a concise report on the synthesis pathways of biolubricants using transition metal complexes (homogeneous) and supported metal complexes (heterogeneous) as efficient catalysts.

5.2 Synthesis of Biolubricants

5.2.1 Esterification

The general strategy for ester-based biolubricant synthesis is to react polyols with long-chain fatty acids in the presence of a catalyst. Varying the polyhydric alcohol manifests in the lubricant properties like viscosity, freezing point, pour point, and flash point. Similarly, the fatty acid chain length determines the viscosity and melting point of the resulting lubricant [17]. Biolubricants with longer carbon chains have higher melting points and greater viscosity, whereas biolubricants with higher double bonds have lower melting temperatures and viscosity, as well as worse oxidative stability. The fatty acid chain for biolubricant production is usually between C12 and C24. As a result, fatty acids and polyol-starting materials can be matched to fine-tune the properties of biolubricants as per specific requirements [18].

5.2.2 Transesterification

Transesterification is the process of converting one ester into another. The transesterification reaction, which involves triglycerides and polyhydric alcohols along with catalysts, is another common method for making ester-based biolubricants. Triglycerides derived from plant-based oils like soybean, palm, and coumarin, are used in the transesterification process in the presence of various Lewis acidic catalysts. The different catalysts utilized in the transesterification process have been examined in this work, particularly the Sn(II) and Sn(IV)-based complexes. The catalytic activity of these catalysts was analyzed with respect to that of H_2SO_4, a common inorganic acid catalyst.

5.2.3 Hydrogenation

As mentioned earlier, the hydrogenation of esters makes the biolubricants suitable for long-term applications. Ester-based lubricants, after hydrogenation, become suitable for various applications in automobile engines and heavy machinery due to the presence of high saturated carbon content. Non-hydrogenated ester-based lubricants generally contain high unsaturation, and so are not useful for many applications as they may lead to undesired side-products. Moreover, such biolubricants are not entirely biodegradable as well.

5.2.4 Simultaneous Hydrogenation–Esterification

Biomass can be converted to biolubricants in a one-step upgradation method of hydrogenation–esterification. Bio-oils derived from biomass consist of several chemicals like carboxylic acids, aldehydes, ketones, and phenols, which degrade the quality of the ester for lubrication purposes. Hence, hydrogenation by H_2/Ni followed by esterification over Lewis acid catalysts transforms all the vulnerable species present in the bio-oil into stable biolubricants.

5.3 Catalysts for Biolubricant Synthesis

5.3.1 Catalysts for Esterification

Lewis-acid catalysts or inorganic acids like HCl generally catalyze the esterification reaction between long-chain carboxylic acids (C4–C14) and polyols [19]. A side effect of inorganic acid catalysis is that the reaction vessels get corroded in the strongly acidic environment, and the salt that is formed by the neutralization reaction during work-up deposit scales on the reaction vessels. In addition, product cleaning and water separation from the crude incur additional costs [20]. Hence, metal complexes as catalysts for the esterification reaction can be beneficial in this regard and is therefore reviewed here as follows:

Oliveira *et al.* [21] used a homogeneous tin complex catalyst, i.e., [Sn(II)(3-hydroxy-2-methyl-4-pyrone)] for poly-esterification of 2,2-Dimethyl-1,3-propandiol, 2,2-bis(hydroxymethyl)butanol and terephthalic acid at 150°C–230°C. The catalytic mechanism is assumed to be based on the Lewis acidic character of Sn(II) along with the ability of the complex to extend the metal d-orbitals, thereby giving access to the substrate via an associative ligand exchange. It was observed that a higher catalyst loading reduced the reaction time, whereas higher yields were observed with low catalyst amount. Noor *et al.* [22] investigated the efficiency of mineral acids as catalysts for the esterification of fatty acid (FA) derived from Jatropha curcas oil with trimethylolpropane (TMP). The reactions were undertaken at 150°C for 3 hours, using an FA:TMP molar ratio of 4:1 and a 2% w/w concentrated catalyst. It was observed that the ester yield was proportional to the acidity of the mineral acid catalysts, where perchloric acid had the highest yield of 70%. Sulfuric acid, p-Toluenesulfonic acid, hydrochloric acid, and nitric acid showed yields of 46%, 42%, 41%, and 37%, respectively [22].

Abiney *et al.* [23] investigated the catalytic activity of $SnCl_2$ in the conversion of oleic acid to ethyl oleate in an ethanol solution, using a 100:1 (v/w) oleic acid:catalyst ratio and a temperature of 80°C for 6 hours of reflux. $SnCl_2$ offers more advantages than mineral acid catalysts on account of its cost-effectiveness, lesser corrosiveness, and Lewis-acid properties. $SnCl_2$ is used only in small quantities as the catalyst, and therefore eliminates the tiresome and wasteful step of pH neutralization of the end products. $SnCl_2$ is a more efficient catalyst compared to mineral H_2SO_4 acid, as well as possesses greater selectivity and a higher conversion rate. The quantity of catalyst utilized and the reaction time determine the conversion rate and selectivity. Ferric salts, in particular $FeCl_3.6H_2O$, were the most effective, with reported yields of 90% in this instance.

Ni(II) and Co(II) mixed ligand complexes with 8-hydroxyquinoline as the primary ligand and amino acids as the spectator ligands were utilized for ester hydrolysis by Shivankar and co-workers [24]. To investigate the reaction process and kinetics, the substrate/catalyst ratio was maintained at 100–500 (v/w), and the amount of ester produced at a temperature of 40°C–50°C was recorded. Five milliliters of reaction mixture was extracted and titrated against a standardized NaOH solution at regular intervals to determine the acid concentration. Mixed ligand metal complexes were shown to be more effective catalysts for ester hydrolysis in this experiment. Shivankar *et al.* [25] also studied the effects of carbohydrates as spectator ligands for ester hydrolysis.

Sn(IV)-based compounds were used as catalysts for esterification by Meneghetti *et al.* [26]. The idea was to utilize the unoccupied d-orbitals of Sn by coordination with the starting materials like alcohol, which have non-bonded electron pairs. To begin, Sn(IV) complexes formed an intermediate product by bonding with alcohol. Further reaction of this intermediate with the carboxylic acid gives the ester and water. The esterification reactions were also catalyzed by hybrids like L-phenylalanine–Ru(III) complex/Si, as reported by Adam *et al.* [27]. The reaction was started with 0.10 g catalyst and a 1: 3 C_2H_5OH: CH_3COOH at 85°C, with a 34% conversion in the first hour and a 94% transformation over the next 12 hours. The complex catalyst that became acidic due to silica grafting onto the Ru(III) site efficiently speeds up the esterification process. Moreover, the catalyst is fairly recyclable after treatment with ethanol. Biolubricants were made utilizing three-dimensional titanosilicate compounds, Ti-SBA-12 and Ti-SBA-16, by Kotwal *et al.* [16]. Esterification was proposed to take place at the Ti sites, which are Lewis acidic in nature. Equal parts of oleic acid and polyol (TMP, NPG, and PE) are the starting materials in a typical reaction along with 0.12 g of catalyst. The reaction mixture was refluxed for 1 hour at 180°C.

5.3.2 Catalysts for Transesterification Reaction

The transesterification (ester-to-ester transformation) process is another common way to make ester-based biolubricants. In comparison to normal acid-catalyzed esterification methods from carboxylic acids and alcohols, transesterification is more favorable. Some carboxylic acids, for example, are only sparsely soluble in organic solvents and have a tough time homogenizing esterification, whereas esters are commonly soluble in most organic solvents. For primary carboxylic acids that are difficult to isolate, the ester-to-ester transition is beneficial. Sodium and potassium alkoxides are the most common bases. Alkali-catalyzed transesterification is faster than acid-catalyzed and occurs at lower temperatures than the latter. However, acid-catalyzed transesterifications are favored because acid catalysts remain unaffected by free fatty acids in the starting material. In practice, acid catalysts can catalyze both the esterification and transesterification reactions at the same time. Methanol is the most widely used alcohol for this purpose, and the process is also known as methanolysis.

Hao *et al.* undertook transesterification of methyl butanoate and n-octanol, catalyzed by Sn(IV), Hf(IV), and Yb(III), Sn(IV) and Hf(IV) complexes of perfluorooctanesulfonylamide [28]. These complexes are essentially Lewis acidic in nature. The $Sn[N(SO_2C_8F_{17})_2]_4$ catalyst demonstrates a good yield of 89% with 99% selectivity for transesterification with equimolar starting materials in a fluorous biphasic solution. The catalytic action of Hf(IV), Sn(IV), and Yb(III) complexes was tested using a fluorous solvent as it improves catalyst immobilization and phase separation. The efficiencies of the metal complex catalysts were tested at 80°C for 15 hours, and yields reported are tabulated (Table 5.1). $Sn[N(SO_2C_8F_{17})_2]_4$ was found to be the outperforming catalyst.

Abreu *et al.* [29] investigated three homogeneous catalysts for the transesterification of triglycerides with methanol: (i) $Sn(HMP)_2(H_2O)_2$; (ii) $Pb(HMP)_2(H_2O)_2$; and (iii) $Zn(HMP)_2(H_2O)_2$ (HMP=3-hydroxy-2-methyl-4-pyrone). Methanolysis of vegetable oils was carried out at 60°C with 400: 100: 1 (v/v/w) methanol:oil:catalyst for 1 hour. The catalytic performances of these complexes are summarized in Table 5.2. The study found that catalytic activity declines in the order Sn^{2+} complex >Zn^{2+} complex >Pb^{2+} complex, concomitant with a reduction in Lewis acidic character.

Table 5.1 Catalyst performance for transesterification reaction between methyl butanoate and *n*-octanol.

Catalyst	Yield (%)
$Sn[N(SO_2C_8F_{17})_2]_4$	89
$Sn[N(SO_2C_8F_{17})_2]_2$	84
$Sn(OSO_2CF_3)_2$	83
$Hf[N(SO_2C_8F_{17})_2]_4$	76
Hf(OSO CF)	65
$Yb(OSO_2CF_3)_3$	12

Table 5.2 Catalytic performance of complexes in the methanolysis of various vegetable oils.

Oils	Catalysts	Yield (%)
Soybean	$Zn(HMP)_2(H_2O)_2$	15.50
	$Pb(HMP)_2(H_2O)_2$	4.20
	$Sn(HMP)_2(H_2O)_2$	37.10
Andiroba	$Zn(HMP)_2(H_2O)_2$	11.2
	$Pb(HMP)_2(H_2O)_2$	5.2
	$Sn(HMP)_2(H_2O)_2$	23.3

Serra *et al.* [30] also reported on the use of homogeneous Sn(IV)-based catalysts for the methanolysis of castor and soybean oils. Dibutyltin diacetate ((C4H9)Sn(C2H3O2)2, di-n-butyl-oxo-stannane ((C4H9)2SnO), butylstannoic acid ((C4H9)SnO(OH)), and dibutyltin dilaurate ((C4H9)2Sn(C12H23O2)2 were used as Lewis acid catalysts in transesterification reactions. The reactions were carried out in two different types of reactors: an open glass reactor (OG) with methanol reflux at atmospheric pressure and a closed steel reactor (CS) with temperatures of 80° C and 120° C. In both reactors, the reaction conditions were the same with continuous stirring. The castor oil methanolysis showed lower yields than soybean oil for the same catalyst, indicating that the chemical composition of the starting material influences the process. Another noteworthy finding was the use of a CS reactor rather than an OG reactor, which resulted in higher reaction yields for all Sn(IV) complexes due to the higher reaction temperature.

Ferreira *et al.* [31] studied the transesterification of soybean oil in the presence of $(C_4H_9)SnO(OH)$, $(C_4H_9)_2SnO$, $(C_4H_9)_2SnO$, and $(C_4H_9)_2Sn(C_{12}H_{23}O_2)_2$ catalysts under mild conditions with long reaction time. The $(C_4H_9)_2$ $Sn(C_{12}H_{23}O_2)_2$ catalyst showed the highest catalytic activity with a 43% yield after 10 hours.

The Lewis acidity, according to Meneghetti *et al.* [26], is behind the catalytic performance of Sn(IV) complexes in transesterification processes. Sn(IV) can coordinate with various species in solution and undergo associative exchange with acids, alcohols, and esters to generate an intermediate product in the process. These intermediates disintegrate to yield new esters and alcohols.

Meneghetti *et al.* studied the catalytic activity of Sn(IV) compounds in transesterification reaction using BuSn(O)OH, $Bu_2Sn(Lau)_2$, and Bu_2SnO as catalysts [26]. $Bu_2Sn(Lau)_2$ proves to be the most efficient catalyst owing to its greater solubility compared to its counterparts. $Bu_2Sn(Lau)_2$ produced an outstanding yield (98%) when MeOH:soybean oil:catalyst is 400:100:1 (v/v/w) with continuous stirring at 150°C. Sn(II) and Sn(IV) complexes are regarded as the best catalysts for transesterification reactions in terms of cost and yield according to the above study. Sn possesses a vacant 5d orbital, indicating a Lewis acidic nature. The main advantage of Sn-based complex catalysts is their good solubility in oil, which allows them to coordinate with other electron-donating groups in the organic phase, hence, their wide applicability in transesterification reactions.

Biolubricants or surfactants are widely used in pharmaceuticals, agrochemicals, fine chemicals, and consumer goods. They are typically synthesized by hydrogenating esters. The use of molecular H_2 at higher temperatures and pressures to reduce aldehydes, esters, and ketones for the production of ester-based lubricants is critical to our daily lives. Due to their high oxygen concentration, they are not suited for use in motor lubricants or as fine chemicals because they are not biodegradable. Furthermore, during hydrogenation, the oxygen concentration of ester-based lubricants drops, while the hydrogen content increases; hence, these lubricants have the potential to be employed as engine oils that meet all types of lubrication requirements. Precious metal-based complexes, such as Au, Ag, Pt, Pd, Ir, and In, and those of Ru, Rh, and Os, are promising candidates to catalyze the reduction of carbonyl groups in esters [32, 33]. However, due to the high cost and restricted availability of precious metal-based catalysts, less-priced active metals have been developed. Complexes based on first-row transition metals have recently attracted attention [34]. To date, organometallic complexes have been widely utilized as catalysts for the hydrogenation of esters in order to achieve a homogeneous phase. The resultant ester is substantially hydrogenated, but separating the catalyst from the crude is a formidable process.

Valencia *et al.* [35] investigated the catalytic activity of mixed ligand Co(II) complexes containing BINAP is well known ligand/PPh_3 ligand for the hydrogenation reaction of β-enamino esters. The authors showed [$Co_2(CO)_8$/rac-BINAP] complex system to have increased catalytic activity after 7–15 hours of reaction time 120°C. In addition, the catalytic effectiveness of cobalt complexes for asymmetric hydrogenation was also measured by varying the racemic ligands (Table 5.3).

Chakraborty *et al.* [36] investigated the catalytic activity of iron-based pincer complexes for the hydrogenation of esters to corresponding ester-based compounds. These complexes catalyze the hydrogenation of esters to the corresponding ester-based lubricants when the reaction conditions are maintained at 63 psi/g of H_2 pressure in toluene at 115°C. CE-1270, derived from coconut oil and comprising of methyl myristate (C14, 26%), methyl laurate (C12, 73%), and traces of C10, was hydrogenated in this investigation.

Bio-oils are turned into biolubricants in a one-pot synthetic process with a uniform catalyst and reaction condition in the hydrogenation–esterification

Table 5.3 Effect of chiral ligands on hydrogenation of β-enamino esters (Reaction Temperature: 120°C).

Complex	Ligands	Time (h)	Yield (%)
$Co_2(CO)_8$	Race-BINAP	24	90
	Race-BINAP	15	89
	Race-BINAP	7	90
	(*R*)-Tolu-BINAP	7	92
	(*R*)-H_8-BINAP	7	89
	(*R*,*R*)-DIOP	7	71
	(*R*,*R*)-Met-DuPHOS	7	81

(OHE) process. The end products are hydrogenated by basic processes like esterification [37, 38] and hydrogenation to yield esters, and they become extremely suited for burning hydrocarbons [39, 40]. Fatty acids, phenols, aldehydes, and ketones are the main constituents of bio-oils used in the one-step hydrogenation–esterification process, and they negatively affect the characteristics of bio-oils [41]. Metal-based catalysts like RANEY® Ni (RN) or Lewis-acid catalysts are used in the OHE process to synthesize hydrogenated esters, under solvent conditions (methanol/ethanol). Xu *et al.* [42] investigated a method for converting bio-oils into hydrogenated esters with a Raney-Ni catalyst for the hydrogenation–esterification process. Hydroxyacetone and furfural showed complete transformation in the presence of methanol using Sn/Cu/Mo/Fe-modified Ni catalysts. When no methanol was added, the rate of conversion of CH_3COOH was 35.1%, but with the addition of methanol, it increased to 81.1%. The conversion of methanol reduced from 35.6% to 28.6% when Mo/Raney-Ni was replaced by 5% Fe. In comparison to Mo/Raney-Ni, however, the conversion of phenol was dramatically reduced when Fe was added [43].

Yu *et al.* [44] investigated the use of Al with Pd catalysts to develop bio-oils via the OHE process, starting with furfural and acetic acid as substrates. The OHE studies used equal amounts of furfural and acetic acid (0.10 mol in 10.0-mL toluene), with 0.40 g Pd/C (5%) + 0.40 g Al or 0.40 g Pd/Al (5%) acting as catalysts. The reaction was carried out under ideal conditions of 2.0 MPa H_2 pressure at 150°C temperature, with continuous stirring for 4 hours at 800 rpm. The Pd metal, as well as the Lewis acidic Al sites, had a concomitant effect favoring the OHE reaction.

5.4 Conclusions

The present review suggests that metal complex catalysts could be employed to enhance the yields of biolubricants. The metal complex catalysts showed a high Lewis acidic character. Sn(II) and Sn(IV)-based complexes as catalysts exhibited high efficiency for esterification and transesterification reactions as they show Lewis acidic character, and their vacant 5d orbital participated to associate lone pair containing functional groups like -OH. Moreover, it is seen that transition metal catalysts enhance selectivity, minimizing the formation of unwanted by-products. The development of biolubricants for high-pressure and high-temperature environments is still a challenge in this field of work. Therefore, the importance of developing efficient metal complex catalysts for biomass upgradation is evident and requires much attention.

References

1. Garcés, R., Martínez-Force, E., Salas, J. J., Vegetable oil basestocks for lubricants. *Grasas y aceites*, *62*, 1, 21–28, 2011.
2. Marth, J. S., Renewable lubricants manual: Biobased oils, fluids and greases. *United Bio Lube.*, 2007.
3. Wang, E., Ma, X., Tang, S., Yan, R., Wang, Y., Riley, W. W., Reaney, M. J., Synthesis and oxidative stability of trimethylolpropane fatty acid triester as a biolubricant base oil from waste cooking oil. *Biomass Bioenergy*, *66*, 371–378, 2014.
4. Gao, R., Cao, B., Hu, Y., Feng, Z., Wang, D., Hu, W., Chen, J., Jie, Z., Qiu, H., Xu, K., Xu, X., Lu, H., Zhu, W., Gao, Z., Xiang, N., Shen, Y., He, Z., Gu, Y., Zhang, Z., Yang, Y., Zhao, X., Zhou, L., Li, X., Zou, S., Zhang, Y., Li, X., Yang, L., Guo, J., Dong, J., Li, Q., Dong, L., Zhu, Y., Bai, T., Wang, S., Hao, P., Yang, W., Zhang, Y., Han, J., Yu, H., Li, D., Gao, G.F., Wu, G., Wang, Y., Yuan, Z., Shu, Y., Human infection with a novel avian-origin influenza A (H7N9) virus. *N. Engl. J. Med.*, 368, 20, 1888–97, 2013 May 16, doi: 10.1056/NEJMoa1304459.
5. Nagendramma, P. and Kaul, S., Development of ecofriendly/biodegradable lubricants: An overview. *Renewable Sustainable Energy Rev.*, *16*, 1, 764–774, 2012.
6. Urbanus, J. H., Lobo, R. C., Riley, A. J., European Hazard classification advice for crude oil–derived lubricant base oils compared with the proposed mineral oil mist TLV®. *Applied Occupational and Environmental Hygiene*, *18*, 11, 815–817, 2003.
7. Wie, L., Li, X., Yanxun, L., Tingliang, L., Guoji, L., Process variables of the esterification reaction of 3-pentadecylphenol and acryloyl chloride catalyzed by an ionic liquid. *Chem. Eng. Technol.*, *36*, 4, 559–566, 2013.
8. Stöcker, M., Biofuels and biomass-to-liquid fuels in the biorefinery: Catalytic conversion of lignocellulosic biomass using porous materials. *Angew. Chem. Int. Ed.*, *47*, 48, 9200–9211, 2008.
9. Gallezot, P., Conversion of biomass to selected chemical products. *Chem. Soc. Rev.*, *41*, 4, 1538–1558, 2012.
10. Kumar, A. and Sharma, S., An evaluation of multipurpose oil seed crop for industrial uses (Jatropha curcas L.): A review. *Ind. Crops Prods.*, *28*, 1, 1–10, 2008.
11. Salimon, J., Salih, N., Yousif, E., Biolubricants: Raw materials, chemical modifications and environmental benefits. *Eur. J. Lipid Sci. Technol.*, *112*, 5, 519–530, 2010.
12. Yang, C. Y., Fang, Z., Li, B., Long, Y. F., Review and prospects of Jatropha biodiesel industry in China. *Renewable Sustainable Energy Rev.*, *16*, 4, 2178–2190, 2012.
13. Refaat, A. A., Different techniques for the production of biodiesel from waste vegetable oil. *Int. J. Environ. Sci. Technol.*, *7*, 183–213, 2010.

14. Kumar, R., Kumar, R., Mahiya, K., Mathur, P., Oxidation of substituted benzyl amines using a phenoxo-bridged dimeric nickel (II) complex: Synthesis, crystal structure and catalytic activity. *Transition Met. Chem.*, *40*, 189–195, 2015.
15. Wang, X., Xing, N., Yan, Z. D., Wang, Z. N., Wang, J. X., Bai, F. Y., Xing, Y. H., Preparation, characterization, crystal structures and catalytic oxidation activities of two transition metal complexes of an N-heterocyclic pincer ligand. *Transition Met. Chem.*, *40*, 217–225, 2015.
16. Kotwal, M., Kumar, A., Darbha, S., Three-dimensional, mesoporous titanosilicates as catalysts for producing biodiesel and biolubricants. *J. Mol. Catal. A: Chem.*, *377*, 65–73, 2013.
17. Wagner, H., Luther, R., Mang, T., Lubricant base fluids based on renewable raw materials: Their catalytic manufacture and modification. *Appl. Catal., A*, *221*, 1–2, 429–442, 2001.
18. Nagendramma, P. and Kaul, S., Study of synthetic complex esters as automotive gear lubricants. *J. Synth. Lubr.*, *25*, 4, 131–136, 2008.
19. Ji, H., Wang, B., Zhang, X., Tan, T., Synthesis of levulinic acid-based polyol ester and its influence on tribological behavior as a potential lubricant. *RSC Adv.*, *5*, 122, 100443–100451, 2015.
20. Park, J. Y., Wang, Z. M., Kim, D. K., Lee, J. S., Effects of water on the esterification of free fatty acids by acid catalysts. *Renewable Energy*, *35*, 3, 614–618, 2010.
21. De Oliveira, A. B., Jorge, I. F., Suarez, P. A. Z., De S. Basso, N. R., Einloft, S., Synthesis and characterization of new bivalent tin chelate of 3-hydroxy-2-methyl-4-pyrone and its use as catalyst for polyesterification. *Polym. Bull.*, *45*, 341–344, 2000.
22. Arbain, N. H. and Salimon, J., The effects of various acid catalyst on the esterification of jatropha curcas oil based trimethylolpropane ester as biolubricant base stock. *J. Chem.*, *8*, S33–S40, 2011.
23. Cardoso, A. L., Neves, S. C. G., Da Silva, M. J., Esterification of oleic acid for biodiesel production catalyzed by SnCl2: A kinetic investigation. *Energies*, 1, 2, 79–92, 2008.
24. Shivankar, V. S., Gavali, L. V., Yadav, S. P., Thakkar, N. V., Mixed ligand co (II) complexes: Use as catalysts in the hydrolysis of esters. *IOSR J. Appl. Chem.*, 3, 31–39, 2013.
25. Shivankar, V. and Gavali, L. Chiral mixed ligand complexes as catalysts in the hydrolysis of esters. *Chem. Sci. Trans.*, 2, 4, 1204–1213, 2013.
26. Meneghetti, M. R. and Meneghetti, S. M. P., Sn (IV)-based organometallics as catalysts for the production of fatty acid alkyl esters. *Catal. Sci. Technol.*, *5*, 2, 765–771, 2015.
27. Adam, F., Hello, K. M., Chai, S. J., The heterogenization of l-phenylalanine–Ru (III) complex and its application as catalyst in esterification of ethyl alcohol with acetic acid. *Chem. Eng. Res. Des.*, *90*, 5, 633–642, 2012.

28. Hao, X., Yoshida, A., Nishikido, J., Recyclable and selective Lewis acid catalysts for transesterification and direct esterification in a fluorous biphase system: Tin (IV) and hafnium (IV) bis (perfluorooctanesulfonyl) amide complexes. *Tetrahedron Letters*, 4, 4, 781–785, 2004.
29. Abreu, F. R., Lima, D. G., Hamú, E. H., Wolf, C., Suarez, P. A., Utilization of metal complexes as catalysts in the transesterification of Brazilian vegetable oils with different alcohols. *J. Mol. Catal. A: Chem.*, 209, 1–2, 29–33, 2004.
30. Serra, T. M., De Mendonca, D. R., Da Silva, J. P., Meneghetti, M. R., Meneghetti, S. M. P., Comparison of soybean oil and castor oil methanolysis in the presence of tin (IV) complexes. *Fuel*, 90, 6, 2203–2206, 2011.
31. Ferreira, D. A., Meneghetti, M. R., Meneghetti, S. M., Wolf, C. R., Methanolysis of soybean oil in the presence of tin (IV) complexes. *Appl. Catal., A, 317*, 1, 58–61, 2007.
32. Schaaf, T. and Greven, H., Hydrogenation to products with IV below 1. *Lipid Technology*, 22, 2, 31–35, 2010.
33. Pozdeev, V. A., Safronov, S. P., Levanova, S. V., Krasnykh, E. L., Catalytic hydrogenation of fatty acid methyl esters. *Russ. J. Appl. Chem.*, 85, 261–266, 2012.
34. Zell, T. and Milstein, D., Hydrogenation and dehydrogenation iron pincer catalysts capable of metal–ligand cooperation by aromatization/dearomatization. *Acc. Chem. Res., 48*, 7, 1979–1994, 2015.
35. Alcón, M. J., Corma, A., Iglesias, M., Sánchez, F., From homogeneous to heterogeneous catalysis: Zeolite supported metal complexes with C2-multidentate nitrogen ligands. Application as catalysts for olefin hydrogenation and cyclopropanation reactions. *J. Organomet. Chem., 655*, 1–2, 134–145, 2002.
36. Amézquita-Valencia, M., Cabrera, A., Cobalt-catalyzed hydrogenation of β-enamino esters using an internal mixture of bidentate and monodentate ligands. *J. Organomet. Chem.*, 768, 145–150, 2014.
37. Chakraborty, S., Dai, H., Bhattacharya, P., Fairweather, N. T., Gibson, M. S., Krause, J. A., Guan, H., Iron-based catalysts for the hydrogenation of esters to alcohols. *JACS*, 136, 22, 7869–7872, 2014.
38. Xiong, W. M., Zhu, M. Z., Deng, L., Fu, Y., Guo, Q. X., Esterification of organic acid in bio-oil using acidic ionic liquid catalysts. *Energy Fuels, 23*, 4, 2278–2283, 2009.
39. Zhang, S., Yan, Y., Li, T., Ren, Z., Upgrading of liquid fuel from the pyrolysis of biomass. *Bioresour. Technol.*, 96, 5, 545–550, 2005.
40. Zhang, B., Lin, L., Zhuang, J., Liu, Y., Peng, L., Jiang, L., Hydrogenation of ethyl acetate to ethanol over Ni-based catalysts obtained from Ni/Al hydrotalcite-like compounds. *Molecules*, 15, 8, 5139–5152, 2010.
41. Kuwahara, Y., Magatani, Y., Yamashita, H., Ru nanoparticles confined in Zr-containing spherical mesoporous silica containers for hydrogenation of levulinic acid and its esters into γ-valerolactone at ambient conditions. *Catalysis Today*, 258, 262–269, 2015.

42. Yang, X., Chatterjee, S., Zhang, Z., Zhu, X., Pittman Jr, C. U., Reactions of phenol, water, acetic acid, methanol, and 2-hydroxymethylfuran with olefins as models for bio-oil upgrading. *Ind. Eng. Chem. Res.*, *49*, 5, 2003–2013, 2010.
43. Xu, Y., Zhang, L., Chang, J., Zhang, X., Ma, L., Wang, T., Zhang, Q., One step hydrogenation–esterification of model compounds and bio-oil to alcohols and esters over Raney Ni catalysts. *Energy Convers. Manage.*, 108, 78–84, 2016.
44. Yu, W., Tang, Y., Mo, L., Chen, P., Lou, H., Zheng, X., Bifunctional Pd/Al-SBA-15 catalyzed one-step hydrogenation–esterification of furfural and acetic acid: A model reaction for catalytic upgrading of bio-oil. *Catal. Commun.*, *13*, 1, 35–39, 2011.

6

Effect of Integration of Nanostructured Semimetals on Lubrication Performance of Non-Edible Vegetable Oil-Based Biolubricants

Umar Farooq[1*] **and Farha Naaz**[2]

[1]*Department of Chemistry, School of Basic and Applied Sciences, Galgotias University, Greater Noida, India*
[2]*Nanochemistry Laboratory, Department of Chemistry, Jamia Millia Islamia, New Delhi, India*

Abstract
Lubricants, as anti-friction media, control friction and wear of components in contact during the motion of different parts of machines. The effective lubrication at desired conditions is required for a consistent and safe operation of an automotive. Current and expected prospects of mineral oils in lubrication have been forecasted as bleak in the near future. So, identifying a promising alternative to mineral oil-based lubricants is needed. With increasing concerns for energy conservation, environmental pollution, and emission reduction, researchers are looking for new ways to develop eco-friendly lubricants. Due to its non-toxicity, biodegradability, and zero greenhouse gas emissions, non-edible vegetable oils can become promising biolubricant alternatives. Non-edible vegetable oils-based lubricants have many valuable physicochemical properties such as high lubricity, high viscosity index, and minimum evaporative losses during transportation, which provide numerous technical advantages over mineral oils. Using nano-additives can effectively enhance the properties of biolubricants thereby expanding their market potential. This chapter illustrates the integration of nanostructured semimetals-based compounds with non-edible vegetable oils for lubricating applications. Additionally, the tribological properties of these nano-additives will be discussed. Based on nanosized semimetal-based compounds,

**Corresponding author*: darumer27@gmail.com; umar.farooq@galgotiasuniversity.edu.in

Mohd Yusuf, Lalit Prasad and Shafat Ahmad Khan (eds.) Nanolubricants: Generation and Applications, (149–172) © 2024 Scrivener Publishing LLC

nanolubricant additives can be categorized as boron-based, silica-based, or carbon allotropes-based additives. Moreover, the lubrication mechanisms, challenges, and prospects of these nanolubricant additives will be explained systematically, which helps in the design and synthesis of novel additives with significant lubrication and anti-wear properties.

Keywords: Lubrication, wear, friction, nanolubricants, biolubricants, coefficient of friction, tribology

6.1 Introduction

Depletion of fossil fuel resources leads to an increase in expenses on petroleum products. In this decade, researchers are extensively focusing on the advancement of renewable resource-derived eco-friendly lubricants. Environmental protection against the pollution caused by petroleum-based lubricants requires a sincere effort. Approximately 50% of all globally used lubricants ends up into the environment due to factors such as instability, spills, or applications leading to total loss. Such environmental threats can be eluded by preventing abominable losses, retrieving, and recycling mineral oil-based lubricants, or utilizing environmentally friendly lubricants [1, 2]. Vegetable oils-based lubricants offer outstanding tribological properties, flash points, and high viscosity indices as these vegetable oils are easily biodegradable and thus, are considered as remarkable candidates as base fluids in eco-friendly lubricants [3]. Contrary to mineral oil-based lubricants, vegetable oil exists with some of their inherent drawbacks including low-temperature stability and poor oxidation, which must be rectified with slight modifications. The preliminary method to address such problems is chemical modifications at the unsaturation sites of vegetable oils [4]. Non-edible vegetable oils are efficient candidates as biolubricants, as they are non-hazardous and biodegradable; they do not emit any toxic gases on their application, which make them environment-friendly. Since ancient times, vegetable oils have been widely employed in lubrication for minimizing friction and wear, though the properties of natural oils have attained more consideration among researchers from the early 20th century [5, 6]. Nowadays, to reduce the utilization of petroleum-based oils in lubrication due to perturbation of depletion of crude oil reserves, environment preservation, and increments in costs, the research related to natural oils has been marked up [7, 8]. As per reported studies, in comparison to synthetic oils, biolubricants consisting of non-edible vegetable oils possess higher lubricity, high viscosity, lower volatility, higher shear stability and higher load-carrying potential. The composition of triacylglycerol and long chains of polar fatty acids is one of the main reasons for turning towards biolubricants [9, 10]. Such constituents of vegetable oils are expedient for boundary lubrication regime due to their close-packed nature, sticking ability on metal surfaces and development of a monolayer, which is helpful in friction and wear reduction [11]. Apart from the above-mentioned properties of bio-based lubricants, it has been observed that they bear higher flash point, higher power output, minor loss of friction, and enhanced efficiency as compared to petroleum-based oils, which make them a suitable alternative to petroleum-based oils [12]. Utilization of non-edible oils for bio-based lubricants could be sustainable, as most of such crops are cultivated in rough environments and do not occupy lands for edible-oil harvest. They incur lower cost, grow easily, and have the potential to generate earnings from bio-based lubricants [13]. Inedible oils are obtained from the crops like neem, mahua, karanja, rubber seed oil, Polanga, Jatropa curcus, castor, jojoba, mongongo, hingan, coriander, nahor, linseed, *Nicotina tabacum*, *Crambe abyssinica*, *Euphorbia tirucalli*, *Acrocomia aculeata*, *Sapindus mukorossi*, and *Sapium sebiferum* [1]. Besides the benefits of non-edible vegetable oil-based lubricants, there are also some disadvantages such as solidification at low temperature, bacterial deterioration, and low thermal-oxidation stability, which can be improved by the incorporation of some additives available as per their applications [14]. Several nano- and micro-sized particle additives were included into bio-based lubricants to enhance the proficiency and performance of lubrication [15, 16]. Solid nanoparticles exhibit essential characteristics for enhanced lubrication properties as they offer a lower coefficient of friction and lower wear, carrying potential to shear between the interacting surfaces and their inherent layered structures [17]. Nanoparticles within the range of 1 nm to 100 nm offer specific characteristics, which are not exhibited in the same material at macroscopic dimensions. These properties were broadly studied in various fields including optics, electronics, medicine, and biotechnology. Dispersion of nanosized particles into a lubricant oil results in nanoscale colloidal suspension, generally termed as a 'nanolubricant' [18]. Nanolubricants exhibit enhanced tribological properties as compared to the base lubricant, which must be because of the combination of certain features of nanoparticles. Nanoparticles can act as spacers, rolling in the middle of friction surfaces and modifying the pure sliding friction to mixed sliding–rolling friction. They also reduce compression stress concentrations related with high contact pressure by bleakly holding the compressive force. Due to their smaller size, nanoparticles can fill the surface roughness points, which provide a mending effect and during the frictional process, form a thin tribofilm that separates the rubbing faces and bears the load. Moreover, because of high surface area/volume ratio, nanoparticles offer a large interfacial surface area to enhance interaction with the surface material of friction pairs forming a protective film, thus reducing friction and wear [19–21]. Numerous nanoparticles have been extensively employed as lubricant additives including organic materials such as graphite, Teflon, graphenes, fullerenes, and diamonds, and inorganic materials like metals, metal oxides, sulfur, hydroxides, borates, and silicates [22, 23]. Among them, semimetal-based nanoparticles, i.e., boron and silica (B, Si), are attracting the attention of researchers due to their superior dispersibility and stability in many types of organic solvents and unique combination of tribological properties [18, 24–28].

This chapter encompasses a thorough assessment of non-edible vegetable oil-driven biolubricants to determine and evaluate their properties, characteristics, advantages, and applications to employ them as alternatives to petroleum oil-based lubricants. This chapter also provides a detailed investigation on the incorporation of semimetal (boron, silica)based nanoparticles and additives into non-edible vegetable oil-based lubricants to improve the tribological properties, as compared to base lubricant due to the result of combinations of significant features of such nanomaterials.

6.2 Lubrication and Lubricating Materials

The science of interacting surfaces or two moving bodies in relative motion is studied and learned with the study of tribology. The association of tribological studies is with wear, friction, lubrication, corrosion, degradation of metal surfaces, engine life, and energy losses. However, this phenomenon can bring about extremely high energy losses caused by friction and wear, which should be minimized. Lubrication is essential to overcome and reduce friction between two interacting parts. For a tribological system, the mechanical properties depend on the hardness of the material, surface toughness, contact geometry and the sliding mechanism of the rubbing part as depicted in Figure 6.1 (a). As per the Stribeck curve, there are three acclaimed lubrication regimes, namely, boundary lubrication, elastohydrodynamic lubrication, and hydrodynamic lubrication as shown in Figure 6.1 (b) [29]. The modern period of lubrication began with the research work of Osborne Reynolds (1842–1912), which was concerned with shafts rotating in bearings [30]. Lubrication results from the separation of two surfaces by a lubricating film. The process of lubrication comprises

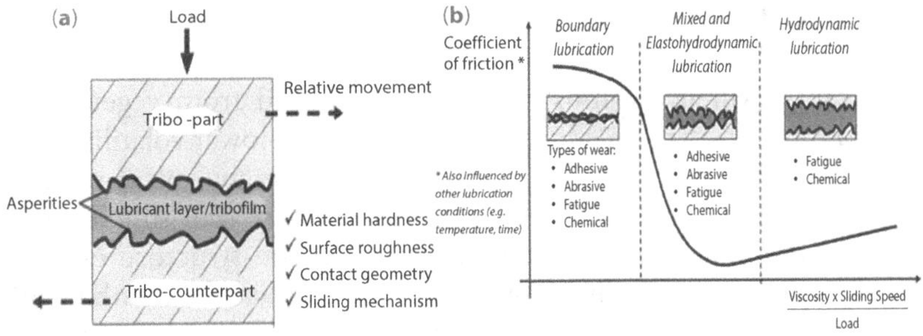

Figure 6.1 (a) Mechanical properties of a tribological system and (b) types of lubrication regimes and conditions [29].

many functions including reduction of wear and heat loss, which occur due to the contact of surface in motion, reduction of oxidation to prevent from rusting, and performance as closure against water, dust and dirt, and as insulator in transformer applications. The substance, which reduces wear and fiction by offering a protective film between two moving surfaces, is termed as a lubricant, which is accessible in liquid, solid, and gaseous forms. The characteristics of an ideal lubricant comprise high resistance to oxidation, high viscosity index, high boiling point, thermal stability, low freezing point, and corrosion prevention capability [31].

The global need for the development of an alternative lubricant to prevent environmental regress has been even more pronounced given the depletion of crude oil sources and a growing spike in prices of petroleum-based oils. Therefore, a reliable lubrication system is essential at optimal operating conditions to inimizes friction and wear, making machines more secure, economical, and reliable to operate. Oils derived from minerals have been used for decades to lubricate automobiles. As a distilled by-product of crude oil, its usefulness is constrained by the availability of crude oil reservoirs. Burning mineral oil causes concern for the environment because it releases trace amounts of metals like iron, zinc, calcium, and magnesium. Additionally, both land and sea ecosystems are vulnerable to the devastation caused by mineral oil disposal. As such, the outlook for mineral oil-based lubricants in automobiles, both now and in the future, is not seen as very promising. It is therefore necessary to find a fluid that may replace mineral oils while still providing the desired tribological qualities [32].

6.3 Inedible Vegetable Oils-Based Biolubricants

Biolubricants, which are mostly lubricants derived from vegetable oils, demonstrate good attributes such as better lubricity, lower volatility, higher flash point, and higher viscosity index. Because vegetable oils contain long chains of fatty acids as well as polar groups, they are well-suited for both hydrodynamic and boundary lubrication [33]. The structure of triacylglycerides consists primarily of long chains of hydrocarbons with terminal functional groups that have carbon, hydrogen, and oxygen atoms. Most vegetable oils contain natural esters. These characteristics distinguish them from the chemical structure of mineral oils, which consists of interconnected chains and rings of carbon and hydrogen. One of the most important qualities, which is completely dissimilar to that of mineral oil, is that the carbons in biolubricants might or might not be saturated with hydrogen. When compared to mineral oils derived from petroleum, the polarity of vegetable oils is higher because of the existence of triglyceride units. As a direct consequence of this, triglycerides have a stronger attraction on the surface of metals. Triacylglycerides, which are also known as triglycerides, are molecules of glycerol that have three esters of long-chain fatty acids connected to each hydroxyl group of the glycerol molecule. Based on chemical structure, the interaction of metal surface with mineral oil-based lubricants differs significantly compared to vegetable oil-based lubricants as shown in Figure 6.2 [34, 35].

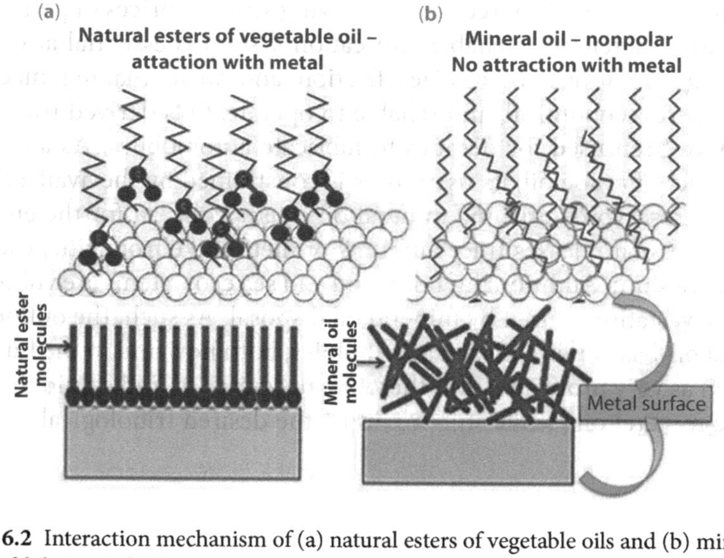

Figure 6.2 Interaction mechanism of (a) natural esters of vegetable oils and (b) mineral oil-based lubricants [34].

Although there are various sustainable feedstock to produce biolubricants, there are limitations on the suitability of edible vegetable oil feedstock due to the latest concerns like their involvement in the human food cycle and environmental degradation caused using available cultivation land. Involvement of edible vegetable oil feedstock could cause ecological imbalances due to the cutting down of forest land for plantation purposes [36]. Therefore, non-edible plant oils are frequently used in the diesel engine sector as biolubricants.

Over 300 oil-producing plant species are grown in various parts of the world. Most non-edible plants may be planted on wastelands to reclaim them; thus, they do not compete with food crops for restricted areas, are relatively affordable, available, and produce equivalent or even larger yields of biolubricants and fuel characteristics than the edibles. The use of non-edible vegetable oil as fuel for engines has been documented by a significant number of researchers; nevertheless, the use of non-edible vegetable-based lubricants in automotive applications has been recorded by only a small number of researchers [37, 38].

Non-edible vegetable oils, also called second-generation fuel sources, can be used to make cutting fluid. Culinary oils are too expensive to use as cutting fluids. Making cutting fluid from non-edible oils avoids the complications of using edible oils. Non-edible oil's potential as a cutting fluid must be studied. Because of its physicochemical features, it is feasible to determine if a vegetable oil is edible or non-edible. Acquiring this information will boost the potential of manufacturing cutting fluids from non-edible vegetable oils [39]. A chemical property is any material attribute that emerges during a chemical reaction, any trait that can only be formed by altering a substance's chemical characteristics. To determine a material's chemical characteristics, its internal composition must be changed. Physical attributes can be measured or evaluated without influencing a material's chemistry. When machining at high temperatures, chemical reactions occur, and the heat creates mist that can be breathed. Machining safety requires knowledge of physicochemical properties. The applicability of vegetable oils in the industry for the generation of biofuels, biolubricants, or polymers depends on their physicochemical properties. Determining the best operating parameters and scaling-up of these processes require knowledge of viscosity, acid value, density, refractive index, specific heat capacity, saponification value, iodine value, and peroxide value. These physicochemical parameters are intrinsic to oils and show their unsaturation [39, 40].

6.3.1 Resources

Non-edible vegetable oils have the potential to replace edible vegetable oils in biolubrication since they can address difficulties with the environment, energy, and the food-versus-fuel conflict. They also have the advantages of being less poisonous and biodegradable. They can also be planted next to highways, irrigation canals, railroads, impoverished areas, degraded woods, and unusable agricultural land. They can advance in growth without fertility and moisture since they are well adapted to desiccate and semi-desiccate environments. Following the extraction of oil from oil-bearing seeds, soil enrichment can be accomplished using seed cakes. Several non-edible plant oils might be viewed as sustainable feedstock for the manufacture of biolubricants, though this may vary from one nation to another [41]. Jatropha curcas, polanga, karanja, neem, mahua, castor, jojoba, coriander, hingan, and mongongo are among the crops that produce non-edible vegetable oil seeds. Other non-edible oils have been utilized as feedstock, including nahor oil, linseed oil, rubber seed oil, *Sapium sebiferum* (Chinese tallow), *Sapindus mukorossi* (soapnut), *Euphorbia tirucalli* (milk bush), *Acrocomia aculeate* (macauba), *Crambeabyssinica* (hochst), and *Nicotiana tabacum* (tobacco) [42].

6.3.2 Properties

Vegetable oils that are not edible are used to make biolubricants, which have favorable and useful physicochemical properties. Compared to petroleum-based lubricants, they offer several technical advantages. They have a high viscosity index, a high flash point, and they suffer from reduced evaporative losses when being stored. They are also very lubricious. The physical characteristics of mineral oils are often dictated by their carbon number distribution, which is determined by the origin of the crude oil. Mineral oils are a complex mixture of straight and branched-chain paraffinic, naphthenic, and aromatic hydrocarbons containing 15 or more carbons. Boiling temperatures typically range from 300°C to 600°C, and specific gravities range from 0.820 for light paraffinic base/process oils to just over 1.0 for strong aromatic base/process oils. They have a wide range of physical characteristics. Studies have shown that bio-based oils derived from non-edible plants are more shear stable, less volatile, more lubricious, and able to bear more weight than synthetic oils. They also have a higher viscosity index. They can therefore be used as a good alternative to petroleum-based oils. Bio-based oils are currently gaining popularity due to their triacylglycerol and lengthy chains of polar fatty acid content. Certain bio-based oil compositions are suitable for boundary lubrication regimes due to their compact structure, ability to stick to metallic surfaces, and capacity to form a monolayer that is helpful for reducing friction and wear. The effects of the various fatty acids in the oil, whether used as additives in combination with synthetic oils or as pure lubricants, have been the subject of numerous studies on biolubricants made from vegetable oils like canola oil. When compared to petroleum-based oils, vegetable-based oils have better dispersancy, a higher flash point, less friction loss, more efficiency, and a higher power output [1, 17].

6.3.3 Merits and Demerits of Vegetable Oil-Based Lubricants

Despite their usefulness, vegetable-based oils have a few drawbacks. These include, among others, susceptibility to bacterial spoilage, solidification at low temperatures, and poor thermal-oxidation stability. Due to the poisonous nature, higher expense, and environmental pollution of petroleum-based lubricants, there has been a revived interest in vegetable-based oils in recent years. Used petroleum-based oils can seep back to the planet's surface if they leak or spill. There is also lack of biodegradability in these oils. Thus, getting rid of the used oils is a difficult task. Research on a wide range of oils have revealed that the biodegradability threshold ranges from about 30% to 65% for petroleum-based oils and from about 95% to 100% for bio-based oils. Researchers are putting in a lot of time and effort to discover a solution in renewable and more ecologically friendly, sustainable, and bio-friendly lubricants to help with these degradability issues. When it comes to lubricating industrial materials and machinery, biolubricants derived from oils that are not fit for human consumption offer a mixed bag of pros and negatives. Their advantages include being safe for humans, animals, and the environment. They have a high viscosity index and a high flash point, making them safe for use on factory floors and extending the life of machinery. Non-edible vegetable oil, for instance, has a viscosity index of around 220, while most mineral oils are in the 90–100 range. Moreover, their high viscosity reduces operating temperature, which in turn leads to an average 16% energy savings. These biolubricants are more heat-intensive, containing less sulfur, and they break down in natural environments. Lubricants for industrial and transportation uses are increasingly being made from non-edible vegetable oils, a renewable resource. Products made from vegetable oils are not only safer for the environment, but their disposal is less of an issue as well. When it comes to basic needs as lubricant base stocks, oleochemical esters are a product category that delivers. These esters stabilize the thermal and cold-flow instability of the pristine, non-edible vegetable oils. Polyol esters like trimethyl propane (TMP), penta-erythritol (PE), and neopentylpolyol (NPG) are utilized to make biolubricants [13]. These bio-based esters are ideal for less extreme applications due to their good low-temperature fluidity; nevertheless, they cannot be employed at higher temperatures. The potential for energy independence is just one of the many reasons why biofuels have been widely adopted around the world. Now, biolubricants, which are made from renewable non-edible vegetable oils, are likely to experience comparable growth for a variety of niche uses. More recently, scientists have been inspired to create process technologies for the commercialization of biolubricants derived from oils that are not suitable for human consumption. Esters of heavy alcohols generated from vegetable oil-based feedstock and biolubricants have lubricating qualities comparable to those of mineral oil-based lubricants. Industries are spending in R&D to improve oil recovery from seeds, lower the costs of processes, and discover new markets for biolubricants, even though their prices are double those of traditional petroleum lubricants [31, 43].

As for the drawbacks, biolubricants based on non-edible oils have poor oxidation stability in their raw state. Although, there are significant drawbacks to using biolubricants made from non-edible vegetable oils, including poor storage, expensive feedstock, low pour points, slow evaporation rates, low heating values, and poor cold flow qualities when used in colder climates. Most of these oils are triglycerides, so three hydroxyl groups from the glycerol fragment are esterified with fatty acid carboxyl groups. While an excess of certain poly-unsaturated fatty acids results in unfavorable oxidation behavior, an excess of saturated long chain fatty acids causes bad low temperature behavior. The low temperature behavior is made worse by even extremely long monounsaturated fatty acids. These esters have a high natural viscosity (and viscosity index) due to the triglyceride structure, which also contributes to their structural stability across reasonable working temperature ranges. Because of the high flash point's correlation with a very low vapor pressure and volatility, possible dangers during use are diminished or eliminated. Even yet, they are less resistant to oxidation than mineral oils, and they can generally not tolerate reservoir temperatures beyond 80°C. However, this can be prevented by using the right antioxidants. In comparison to comparable mineral oils, these oils have reduced filterability, greater froth, and less hydrolytic stability. They clog filters, have a disagreeable smell, and are incompatible with paints and surface treatments [44]. By employing N-Phenylalpha-naphthylamine as an antioxidant and partially adding additives, oxidation stability and low pour point were improved. Additionally, transesterification or epoxidation are the methods to improve low-temperature oxidation stability. The small range of viscosities of vegetable oil-based lubricants must be improved in order to make them sustainable. Viscosity, which functions as a protective coating between the surfaces in contact to shield them from wear, is one of the important elements in influencing the coefficient of friction between the sliding surfaces. Viscosity modifiers, which are kind to the environment, can be employed for this. Ethylene-vinyl acetate (EVA) copolymers and oleogels, a traditional bio-based lubricant, have been created. It was discovered that EVA works well as a thickening agent to transform vegetable oil into a bio-based lubricant. The use of ethylene-vinyl acetate and styrene-butadiene-styrene copolymers, which increase kinematic viscosities at 40°C and 100°C, can also raise the viscosity of a bio-based lubricant [31].

6.4 Nanoparticle Additives to Enhance Tribological Performance of Non-Edible Vegetable Oil Lubricants

Small weight percentages of additives are added to the lubricant base stock to improve and enhance the oil characteristics, which help to further reduce wear and friction in the system. These additives might be dispersant, anti-corrosion, anti-wear (AW), and extreme pressure (EP) additives. AW and EP additives are particularly utilized to lubricate mechanical parts. It should be noted that the classic EP and AW additives—chlorine and phosphorus compounds—have had their use restricted for environmental reasons. Researchers are looking into the design of nanoparticles as a new class of lubricant additives as a means of resolving this issue and the performance bottlenecks that have been reached by most currently available lubricants. Since they do not need tribo-active substances like chlorine, phosphorus, and sulfur, which are bad for the environment, most NPs are ecologically friendly and demonstrate improved tribological properties in lubricants. Additionally, nanoparticles in the nanoscale size range can fill contact asperities. They serve as friction modifiers, AW and EP additives, achieve excellent heat stability, and respond to the friction surface instantly. So far, numerous studies have shown the usage of different types of NPs as lubricant additives. Excellent wear and friction reduction is demonstrated by metal, metal sulfides, metal oxide, boron nitrides, carbon materials, nanocomposites, and rare earth compounds [45].

Nanolubricants are lubricant additives made of nanoparticles or nanomaterials. Their diameter particle size typically ranges from 1 to 100 nm. In lab testing, adding nanolubricants to base oils or coatings significantly reduces wear and friction, displaying intriguing tribological features. Both the one-step procedure and the two-step method can be used to create nanolubricants. The nanolubricants are directly developed by a chemical process in the one-step technique. In the two-step process, nanomaterials are first created as dry powders using physical or chemical means, and then they are dispersed into the base oil using mixing methods with or without dispersants or surfactants. Nanoparticles are made up of metal, metal oxide, sulfide, nanocomposites, carbon nanoparticles, and rare earth compounds based on their chemical makeup. Most of the studies conducted and 72% of the reviews are on metal-containing nanoparticles. Studies on carbon nanoparticles, nanocomposites, and rare earth compounds, however, only made up 7%, 6%, and 7% of the total, respectively [46]. A nanolubricant typically consists of three components: the lubricant/base oil solvent; the anti-wear (AW), extreme pressure (EP), or friction modifier (FM) nanoparticles; and the surfactant that lives at the interface between the lubricating oil and the nanoparticles. The usage of nanoparticles as a lubricant component has various benefits. The most crucial characteristic is how small they are, which allows the nanoparticles to enter the contact region and produce a beneficial lubricating effect. Additionally, they are small enough to pass through oil system filters unharmed. The following are the five benefits of using nanoparticles as lubricant additives: (1) insolubility in nonpolar base oils; (2) low reactivity with other lubricant additives; (3) high possibility of film formation on many different types of surfaces; (4) more durability; and (5) high nonvolatility to withstand high temperatures. Nanoparticles are also adaptable, and numerous studies have shown that a single type of nanoparticle can be used as an AW, an EP additive, and an FM. A variety of nanolubricants can be produced by mixing different types, sizes, and morphologies of nanoparticles with lubricants. It is still challenging to choose the right nanoparticle addition, despite numerous studies showing a notable tribological improvement in lubricants dispersed with various types of nanoparticles. The compatibility of nanoparticles with the base oil or lubricant, as well as their sizes, morphologies, and concentrations all show how efficient nanolubricants are. A thorough review is necessary to address all these characteristics while concentrating on their tribological test circumstances and associated lubrication mechanisms to choose the best nanoparticle and lubricant combination [15, 47].

6.4.1 Tribological Performance-Based Categorization of Nanoparticles

Depending on why they are used, distinct types of nanoparticles can be identified. The classification has been given based on the most popular nanoparticles used for tribological performance because this chapter focused on studies pertaining to the tribological performance of nanoparticles. Researchers have created the nanoparticles for the bulk of research investigations, while some other studies have also employed commercially available nanoparticles. Numerous studies conducted over the past few decades have found that adding nanoparticles to lubricants can effectively reduce friction and wear. These nanoparticles include metal, metal oxide, metal sulfides, carbonate, rare earth-based clays, carbon materials, boron-based materials, and silicon-based materials [15, 48]. The form, size, and source of the nanoparticles can all be further differentiated. Due to specific characteristics of the nanoparticles, such as their size, shape, and physicochemical makeup, the friction-reduction and anti-wear behaviors are improved.

6.4.2 Effect of Nanoparticle Dispersion Stability, Shape, Size, Surface, Concentration, and Kind of Tribo-Test on the Tribological Performance

6.4.2.1 Dispersion Stability

The size of the nanoparticles is so small that when they are added to a fluid medium, Brownian motion keeps them dispersed in liquids due to the random movement of the particles. The suspended particles may adhere to one another and create larger agglomerates, which may then gravitationally settle to the ground. In addition to sedimentation, the agglomeration of nanoparticles causes a loss in wear resistance and friction-reducing capability. Dispersion stability is therefore essential for dependable lubrication function. Sedimentation and blockage may happen when there is low dispersion stability. The absence of a considerable rate of particle accumulation is referred to as stability. A stable suspension is regarded as a need for a successful formulation of a nanolubricant [15].

6.4.2.2 Shape of Nanoparticles

Nanoparticles can adopt a variety of different shapes, including those resembling platelets, spherical solids, spherical multilayered particles, and nanoparticles with hollow cores. The contact pressures between the surfaces that are interacting can be used to direct the choice of the shape of the nanoparticle. In most of the research concerning the application of nanolubricants, spherical nanoparticles have been found to play a far more significant role.

6.4.2.3 *Size of Nanoparticles*

Nanoparticles' mechanical and physicochemical properties are affected by their size, which in turn affects their tribological properties. In some regimes, hardness rises as the size of a nanoparticle increases. When nanoparticles that are harder than the shearing surfaces are put between the shearing surfaces, they indent and scratch the surfaces. So, when designing lubrication systems that use nanoparticles, it is important to keep in mind how their size affects their hardness. The ratio of the root mean square roughness of the lubricated surface to the radius of the nanoparticle is an important parameter when choosing the right nanoparticle size. If the nanoparticles are too big compared to the space between the asperities, they will not stick to the contact zone and the lubrication will be poor.

6.4.2.4 *Surface Functionalization*

In nanoparticle-based lubricant formulations, surface functionalization has two different roles: it controls the colloidal stability of nanoparticle dispersion and makes the outermost layers of the nanoparticles more slippery.

6.4.2.5 *Nanoparticles Concentration*

Tribological properties are significantly affected by the concentration of nanoparticles present in base oil. The ideal concentration of nanoparticles is necessary to achieve the desired reduction in friction and wear. Within the context of the tribological enhancement mechanisms, the concentration of nanoparticles also plays a significant role. When excessive levels are present, the collision of hard nanoparticles can produce shearing, which leads to an increase in friction. The sheared nanoparticles have sharper angles, which in turn contribute to an increase in surface roughness and the removal of additional material. When present in very high quantities, nanoparticles are unable to reach the contact region, and so cannot exert their effects.

6.4.2.6 *Nature of Tribo-Testing*

In contrast to dissolved additives, the augmentation of rubbing surfaces for nan1oparticles also heavily depends on the mechanics of the rubbing contact and the presence of separating oil layers. Since the tribological efficiency of nanolubricants is probably quite system-specific, a significantly variable performance is probably going to be seen under various test parameters [15, 49].

6.5 Tribological Mechanisms of Nanoparticles

6.5.1 Ball-Bearing Effect

The spherical nanoparticles transform the friction phenomena from a sliding process to a combination of sliding and rolling friction by acting as nanometric ball bearings that roll between the shearing surfaces.

6.5.2 Protective Film Formation

On the surface, the nanoparticles aggregate into a protective layer that is very thin. Several studies have demonstrated that the creation of a protective film can lessen the amount of friction that occurs, as well as the amount of wear that occurs.

6.5.3 Mending Effect or Self-Healing Effect

Nanoparticles deposit on the contacting surfaces, which help compensate for the loss of mass that occurs in this mechanism. Lubricating oil typically contains nanoparticles, which have the potential to fill up scars and grooves on the friction surface. EDX analysis has been used in most investigations that have reported the mending effect in order to verify the deposition of nanoparticles on the surface that was being rubbed.

6.5.4 Polishing Effect

The polishing effect is thought to be observed when the lubricating surface's roughness is decreased through abrasion with the help of nanoparticles. In some research, the word "polishing effect" also refers to the term "smoothing effect". Less friction and, hence, less wear would result from a reduction in surface roughness [50]. Different mechanism involved in the tribological activity of nanoparticles is given in Figure 6.3.

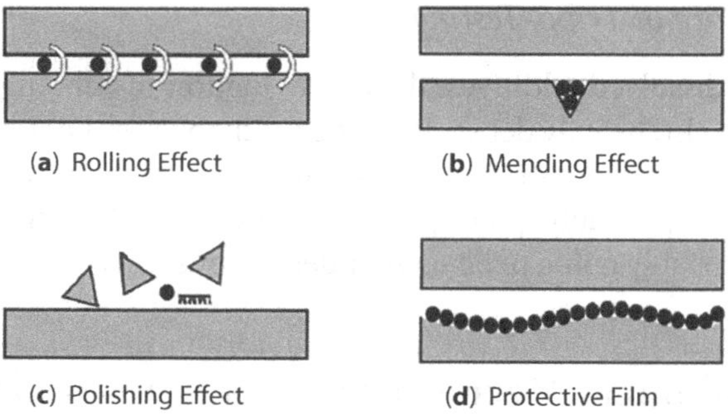

Figure 6.3 Tribological mechanisms of nanoparticles: (a) ball-bearing effect; (b) mending effect or self-healing effect; (c) polishing; and (d) protective film formation [50].

6.5.5 Semimetal-Based Nano-Biolubricants

As discussed earlier, different nanomaterials have been utilized as additives to improve the tribological properties of the base oils of a lubricant. Among these nanodimensional additives, semimetals like boron and silicon-based nanoadditives have been investigated by various researchers for their effect on tribological properties of nano-biolubricants.

6.5.6 Boron-Based Nanoadditives in Non-Edible Vegetable Oils-Based Lubricants

Boron-based nanoparticles have been the subject of research due to the load bearing and anti-wear properties that they exhibit. They are also safe for the environment and are resistant to high temperatures, qualities that make them a credible choice for usage as an additive in lubricating oils. The inclusion of additives with diverse properties, such as anti-wear, anti-friction, and anti-corrosion capabilities, can further improve the creation of lubricating a layer between the contact surfaces. Low friction was provided by solid semimetal-based lubricants like boric acid and hexagonal boron nitride (h-BN), which were in the form of lamellar powder that produced layered crystal formations. This was caused by the fact that h-BN forms strong covalent connections within a layer and weak van der Waals bonds between different layers. The protective boundary lubrication coating that emerged because of the lamellar powder adhering to the contact surface was able to reduce the number of contact surfaces and reduce wear. It tended to align its layers parallel to the direction of motion, which decreased friction [4].

Hexagonal boron nitride (h-BN) is a performance-improving substitute for inorganic solid lubricants like graphite and molybdenum disulfide owing to its special properties. The facile shearing along the basal plane of hexagonal boron nitride's crystalline structure is what gives it its lubricating properties. By mixing boron oxide and ammonia, and grinding the resultant mixture, Celik *et al.* created nanohexagonal boron nitride particles. To the engine oil, various concentrations of micro h-BN particles were introduced. Using a ball-on-disc tribometer, the tribological characteristics of AISI 4140 steel with nanohexagonal boron nitride used as an additive in engine oil, were studied. The viscosity of the lubricants was unaffected by the addition of nanohexagonal boron nitride particles, although the friction coefficients were altered when nano-h-BN was added to engine oil. Oil that contains enough nano-h-BN compounds prevents direct contact, which reduces friction and wear [51]. Reeves *et al.* have investigated the effect of the size of BN over the tribological properties of biolubricants [52]. Canola oil-based lubricant was employed during the investigation. In this study, different sizes ranging from nanometer to submicron to micron of BN were used to investigate the effect of size of BN additives on the lubrication properties of canola-based lubricant oils. During this investigation, wear and friction measurements were carried out using pin-on-disc tribometer under ambient conditions. The change in wear damage and roughness of the surface using different-sized BN samples was investigated by carrying out topographical studies using scanning electron microscopy and an optical profilometer. From this study, it was observed that nanodimensional BN outperformed micron- and submicron-sized BN in terms of wear and friction reduction. The tribological properties like friction and wear due to the addition of nanosized BN in canola oil was reduced by 40% and 70% with a 90% smooth and polished surface. The improved lubrication properties of canola oil due to the addition of nanosized BN was attributed to the small size and spherical particles of BN. These properties provide better coalescence in the asperity valleys compared to micron- and submicron-sized BN, which are larger in size and form plate-like structures. The schematic representation of the influence of size on the formation of lubrication film over the surface is shown in Figure 6.4, which confirms the formation of a smoother surface using nanodimensional BN. From the investigation, it was concluded that the addition of particles smaller than the size of an asperities reduces friction, as well as wear, during the lubrication process. Also, the particle with a larger size reduces the friction of base oils by carrying a portion of load between the asperities, but due to its large size, more abrasions are caused, thus, resulting to high wear and more rougher surfaces.

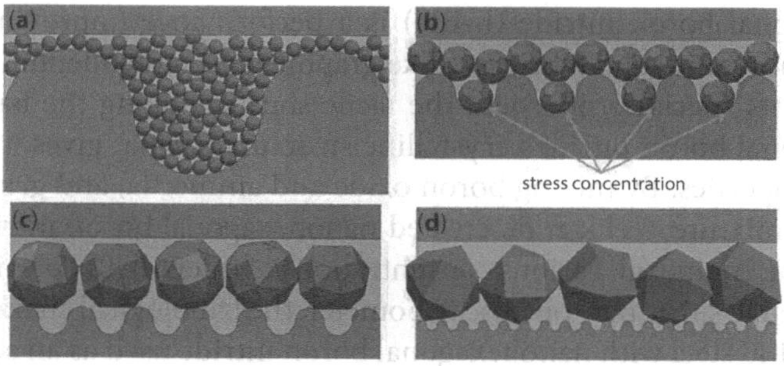

Figure 6.4 Schematic representation of the effect of size on the lubrication mechanism of canola oil-based lubricants at tribo-interface, with sizes ranging from (a) 70 nm, (b) 0.5 mm, (c) 1.5 mm, and (d) 5.0 mm [52].

Similarly, different investigations were carried out to analyze the effect of the weight percentage of nanoadditives on tribological properties of different biolubricants. In one such investigation, h-BN was used to evaluate the effect of its different weight percentage on the tribological properties of polyalphaolefine (PAO). The lubrication parameters were investigated using a pin-on-disc test, and it was observed that the optimum weight concentration of h-BN was equal to 0.5 wt%. Compared to pure PAO, the lubricant developed by the addition of the optimum wt% of h-BN reduced the coefficient of friction by about 5.4%. Similarly, other reports have investigated the synergistic effect of h-BN on the lubrication properties of different mineral oil-based lubricants. From different investigations, the reasons responsible for improved lubrication properties due to addition of h-BN were evaluated. It was deduced that the addition of h-BN either to the vegetable or mineral oil-based lubricants results in the formation of a boron oxide-based tribofilm on the surface of interacting bodies as a result of a tribochemical reaction. Another important factor resulting in the high lubrication performance of h-BN additives was attributed to the conducting nature of h-BN. Due to its conducting properties, h-BN can carry away the heat produced during rubbing of different components of machines working under lubrication conditions, thus improving the stability and viscosity of a lubricant, which in turn reduced the friction and wear of interacting surfaces for a longer period. Similarly, many other semimetal-based nanostructures were used an additive to improve the lubrication properties of non-edible oil-based biolubricants. Among other different materials used to improve the lubrication properties of biolubricants are Si, Sb, and Ge. Recently, a few of the investigations were carried out to analyze the compatibility of Si-based nanoparticles with many biolubricants. Different concentrations of silica nanoparticles were used in

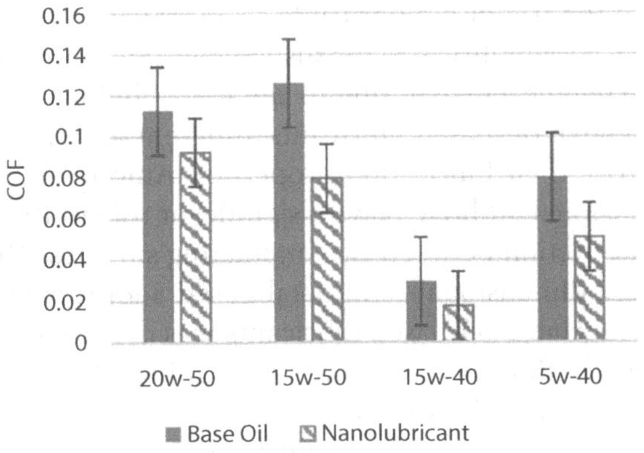

Figure 6.5 Effect of silica nanoparticles on the lubrication properties (friction performance) of synthetic lubricants [53].

configuration with coconut oil. Five different concentrations ranging from 0.25 to 1.25 wt% of SiO_2 nanoparticles were employed during the investigation [53]. It has been observed that optimum concentration with minimum coefficient of friction (COF) was equal to 1 wt% and the COF between interacting surfaces was reduced to 50.85%, compared to pure coconut oil-based lubricant. Figure 6.5 shows the effect of the addition of silica nanoparticles on coefficients of the viscosity of coconut oil. From these investigations, it was concluded that the addition of additives of different sizes significantly affects the tribological properties of vegetable oil-based biolubricants. From these investigations, it was concluded that the addition of nanodimensional semimetal additives in biolubricants showed improved tribological properties to a large extent and the improved lubrication properties were attributed to the small size of the nanoparticles. In addition to the size, a concentration of nanoadditives also plays a significant role in the improved lubrication properties of biolubricants. The addition of an optimum concentration of an additive into a biolubricant is a prime factor in controlling the tribological properties of nano-biolubricants.

6.6 Conclusion

In this chapter, the use of lubricants in reducing friction and wear properties was discussed in the Introduction section. Different characteristics and regions of a lubrication for different lubricants were discussed.

The importance of the use of biolubricants as an crucial replacement for petroleum product-based lubricants for sustainable environment has been described in this chapter. Further, due to the low tribological properties of non-edible vegetable oil-based lubricants, different additives are being employed to improve their lubrication properties. Among different additives, a nanodimensional additive plays a significant role in the reduction of the coefficient of friction between two interacting surfaces. The effect of the addition of nanodimensional semimetal-based additives on tribological properties of vegetable oil-based lubricants was discussed, and it was observed that the addition of these additives significantly improved the tribological properties of the base oil. In addition to the size, the amount of semimetal additives like silica and boron nitride plays an important role in influencing the lubrication properties of base oils. It has been observed that the addition of optimum concentration of these additives has the highest impact on the tribological properties of any lubricant. However, from the literature, it has been observed that not many investigations using nanodimensional semimetal additives in base oils for lubrication have been carried out. Therefore, more explorations of this field of research are needed so that we can fully replace mineral oil-based lubricants with biolubricants for a sustainable future.

References

1. Singh, Y., Sharma, A., Singla, A., Non-edible vegetable oil–based feedstocks capable of bio-lubricant production for automotive sector applications—A review. *Environ. Sci. Pollut. Res.*, 26, 15, 14867–14882, 2019.
2. Cetin, M.H., Kesen, A., Korkmaz, S., Kilincarslan, S.K., Performance evaluation of the nano-silver added vegetable-oil-based cutting fluid in drilling process. *Surf. Topogr. Metrol. Prop.*, 8, 2, 025029, 2020.
3. Shashidhara, Y.M. and Jayaram, S.R., Vegetable oils as a potential cutting fluid—An evolution. *Tribol. Int.*, 43, 5-6, 1073–1081, 2010.
4. Talib, N., Nasir, R.M., Rahim, E.A., Tribological behaviour of modified jatropha oil by mixing hexagonal boron nitride nanoparticles as a bio-based lubricant for machining processes. *J. Cleaner Prod.*, 147, 360–378, 2017.
5. Fox, N.J., Tyrer, B., Stachowiak, G.W., Boundary lubrication performance of free fatty acids in sunflower oil. *Tribol. Lett.*, 16, 4, 275–281, 2004.
6. Lundgren, S.M., Persson, K., Mueller, G., Kronberg, B., Clarke, J., Chtaib, M., Claesson, P.M., Unsaturated fatty acids in alkane solution: Adsorption to steel surfaces. *Langmuir*, 23, 10598–10602, 2007.
7. Sajjadi, B., Raman, A.A.A., Arandiyan, H., A comprehensive review on properties of edible and non-edible vegetable oil-based biodiesel: Composition,

specifications and prediction models. *Renewable Sustainable Energy Rev.*, 63, 62–92, 2016.
8. Nowak, P., Kucharska, K., Kamiński, M., Ecological and health effects of lubricant oils emitted into the environment. *Int. J. Environ. Res. Public Health*, 16, 16, 3002, 2019.
9. Shahabuddin, M., Mofijur, M., Fattah, I.R., Kalam, M.A., Masjuki, H.H., Chowdhury, M.A., Hossain, N., Study on the tribological characteristics of plant oil-based bio-lubricant with automotive liner-piston ring materials. *Curr. Res. Green Sustain. Chem.*, 5, 100262, 2022.
10. Liu, Y. and Binks, B.P., Foams of vegetable oils containing long-chain triglycerides. *J. Colloid Interface Sci.*, 583, 522–534, 2021.
11. Fernández-Silva, S.D., Delgado, M.A., Roman, C., García-Morales, M., Rheological and tribological properties of nanocellulose-based ecolubricants. *Nanomaterials*, 11, 11, 2987, 2021.
12. Panchal, T.M., Patel, A., Chauhan, D.D., Thomas, M., Patel, J.V., A methodological review on bio-lubricants from vegetable oil based resources. *Renewable Sustainable Energy Rev.*, 70, 65–70, 2017.
13. Yang, L., Takase, M., Zhang, M., Zhao, T., Wu, X., Potential non-edible oil feedstock for biodiesel production in Africa: A survey. *Renewable Sustainable Energy Rev.*, 38, 461–477, 2014.
14. Bart, J.C., Gucciardi, E., Cavallaro, S., *Biolubricants: Science and technology*, Elsevier, Netherlands, 2012.
15. Gulzar, M., Masjuki, H.H., Kalam, M.A., Varman, M., Zulkifli, N.W.M., Mufti, R.A., Zahid, R., Tribological performance of nanoparticles as lubricating oil additives. *J. Nanopart. Res.*, 18, 8, 1–25, 2016.
16. Reeves, C.J., Menezes, P.L., Lovell, M.R., Jen, T.C., The size effect of boron nitride particles on the tribological performance of biolubricants for energy conservation and sustainability. *Tribol. Lett.*, 51, 3, 437–452, 2013.
17. Sikdar, S., Rahman, M.H., Menezes, P.L., Synergistic study of solid lubricant nano-additives incorporated in canola oil for enhancing energy efficiency and sustainability. *Sustainability*, 14, 1, 290, 2021.
18. López, T.D.F., González, A.F., Del Reguero, Á., Matos, M., Díaz-García, M.E., Badía-Laíño, R., Engineered silica nanoparticles as additives in lubricant oils. *Sci. Technol. Adv. Mater.*, 16, 5, 055005, 2015.
19. Lee, K., Hwang, Y., Cheong, S., Choi, Y., Kwon, L., Lee, J., Kim, S.H., Understanding the role of nanoparticles in nano-oil lubrication. *Tribol. Lett.*, 35, 2, 127–131, 2009.
20. Liu, G., Li, X., Qin, B., Xing, D., Guo, Y., Fan, R., Investigation of the mending effect and mechanism of copper nano-particles on a tribologically stressed surface. *Tribol. Lett.*, 17, 4, 961–966, 2004.
21. Zhou, J., Yang, J., Zhang, Z., Liu, W., Xue, Q., Study on the structure and tribological properties of surface-modified Cu nanoparticles. *Mater. Res. Bull.*, 34, 9, 1361–1367, 1999.

22. Uflyand, I.E., Zhinzhilo, V.A., Burlakova, V.E., Metal-containing nanomaterials as lubricant additives: State-of-the-art and future development. *Friction*, 7, 2, 93–116, 2019.
23. López, T.D.F., González, A.F., Del Reguero, Á., Matos, M., Díaz-García, M.E., Badía-Laíño, R., Engineered silica nanoparticles as additives in lubricant oils. *Sci. Technol. Adv. Mater.*, 16, 5, 055005, 2015.
24. Baldwin, B.A., Relative antiwear efficiency of boron and sulfur surface species. *Wear*, 45, 3, 345–353, 1977.
25. Liu, W., Jin, Z., Xue, Q., The performance and antiwear mechanism of S-containing organic borate as an oil additive. *Lubr. Sci.*, 7, 1, 49–60, 1994.
26. Klepper, C.C., Williams, J.M., Truhan Jr., J.J., Qu, J., Riester, L., Hazelton, R.C., Moschella, J.J., Blau, P.J., Anderson, J.P., Popoola, O.O., Keitz, M.D., Tribo-mechanical properties of thin boron coatings deposited on polished cobalt alloy surfaces for orthopedic applications. *Thin Solid Films*, 516, 10, 3070–3080, 2008.
27. Li, X., Cao, Z., Zhang, Z., Dang, H., Surface-modification *in situ* of nano-SiO_2 and its structure and tribological properties. *Appl. Surf. Sci.*, 252, 22, 7856–7861, 2006.
28. Kim, D. and Archer, L.A., Nanoscale organic–inorganic hybrid lubricants. *Langmuir*, 27, 6, 3083–3094, 2011.
29. Jason, Y.J.J., How, H.G., Teoh, Y.H., Chuah, H.G., A study on the tribological performance of nanolubricants. *Processes*, 8, 11, 1372, 2020.
30. Ahmed, N.S. and Nassar, A.M., Lubrication and lubricants, in: *Tribology - Fundamentals and Advancements*, J. Gegner (Ed.), IntechOpen, London, 2013, 10.5772/56043.
31. Singh, Y., Farooq, A., Raza, A., Mahmood, M.A., Jain, S., Sustainability of a non-edible vegetable oil based bio-lubricant for automotive applications: A review. *Process Saf. Environ. Prot.*, 111, 701–713, 2017.
32. Ssempebwa, J.C. and Carpenter, D.O., The generation, use and disposal of waste crankcase oil in developing countries: A case for Kampala district, Uganda. *J. Hazard. Mater.*, 161, 2-3, 835–841, 2009.
33. Syahir, A.Z., Zulkifli, N.W.M., Masjuki, H.H., Kalam, M.A., Alabdulkarem, A., Gulzar, M., Khuong, L.S., Harith, M.H., A review on bio-based lubricants and their applications. *J. Cleaner Prod.*, 168, 997–1016, 2017.
34. Bob, F. and Dwight, S., The benefits of bio-based lubricants, 2010, Available in pdf from Gear Solutions: http://www.gearsolutions.com/article/detail/5990/the-benefits-of-bio-based-lubricants.
35. Gnanasekaran, D. and Chavidi, V.P., *Vegetable oil based bio-lubricants and transformer fluids: Applications in power plants*, Springer, United States, 2017.
36. Salih, N. and Salimon, J., A review on eco-friendly green biolubricants from renewable and sustainable plant oil sources. *Biointerface Res. Appl. Chem.*, 11, 5, 13303–13327, 2021.

37. Esonye, C., Onukwuli, O.D., Ofoefule, A.U., Characterization and oxidation modeling of oils from Prunus amygdalus, Dyacrodes edulis and Chrysophyllum albidium. *Ind. Crops Prod.*, 128, 298–307, 2019.
38. Agarwal, A.K., Biofuels (alcohols and biodiesel) applications as fuels for internal combustion engines. *Prog. Energy Combust. Sci.*, 33, 3, 233–271, 2007.
39. Kumar, A. and Sharma, S., Potential non-edible oil resources as bio-diesel feedstock: An Indian perspective. *Renewable Sustainable Energy Rev.*, 15, 4, 1791–1800, 2011.
40. Kazeem, R.A., Fadare, D.A., Ikumapayi, O.M., Akinlabi, S.A., Afolalu, S.A., Akinlabi, E.T., Analysis of the physicochemical properties of some selected non-edible vegetable oil-based cutting fluids using the design of experiment (DOE) approach. *Lubricants*, 10, 2, 16, 2022.
41. Bartolomei, M., Bertocchi, P., Antoniella, E., Rodomonte, A., Physicochemical characterisation and intrinsic dissolution studies of a new hydrate form of diclofenac sodium: Comparison with anhydrous form. *J. Pharm. Biomed. Anal.*, 40, 5, 1105–1113, 2006.
42. Atabani, A.E., Silitonga, A.S., Ong, H.C., Mahlia, T.M.I., Masjuki, H.H., Badruddin, I.A., Fayaz, H., Non-edible vegetable oils: A critical evaluation of oil extraction, fatty acid compositions, biodiesel production, characteristics, engine performance and emissions production. *Renewable Sustainable Energy Rev.*, 18, 211–245, 2013.
43. Changmai, B., Vanlalveni, C., Ingle, A.P., Bhagat, R., Rokhum, L., Widely used catalysts in biodiesel production: A review. *RSC Adv.*, 10, 68, 41625–41679, 2020.
44. Erhan, S.Z. and Asadauskas, S., Lubricant basestocks from vegetable oils. *Ind. Crops Prod.*, 11, 2-3, 277–282, 2000.
45. Hwang, H.S. and Erhan, S.Z., Modification of epoxidized soybean oil for lubricant formulations with improved oxidative stability and low pour point. *J. Am. Oil Chem. Soc.*, 78, 12, 1179–1184, 2001.
46. Singh, Y., Garg, R., Kumar, S., Aspects of non-edible vegetable oil-based bio-lubricants in the automotive sector. *Green*, 5, 1-6, 59–72, 2015.
47. Johansson, J.E., Devlin, M.T., Prakash, B., Lubricant additives for improved pitting performance through a reduction of thin-film friction. *Tribol. Int.*, 80, 122–130, 2014.
48. Jason, Y.J.J., How, H.G., Teoh, Y.H., Chuah, H.G., A study on the tribological performance of nanolubricants. *Processes*, 8, 11, 1372, 2020.
49. Waqas, M., Zahid, R., Bhutta, M.U., Khan, Z.A., Saeed, A., A review of friction performance of lubricants with nano additives. *Materials*, 14, 21, 6310, 2021.
50. Peña-Parás, L., Maldonado-Cortés, D., Taha-Tijerina, J., Eco-friendly nanoparticle additives for lubricants and their tribological characterization, in: *Handbook of Ecomaterials*, Martínez, L., Kharissova, O., Kharisov, B. (Eds.), pp. 3247–3267, 2019.

51. Yan, T., Ingrassia, L.P., Kumar, R., Turos, M., Canestrari, F., Lu, X., Marasteanu, M., Evaluation of graphite nanoplatelets influence on the lubrication properties of asphalt binders. *Materials*, 13, 3, 772, 2020.
52. Reeves, C.J., Menezes, P.L., Lovell, M.R., Jen, T.C., The size effect of boron nitride particles on the tribological performance of biolubricants for energy conservation and sustainability. *Tribol. Lett.*, 51, 3, 437–452, 2013.
53. Xie, H., Jiang, B., He, J., Xia, X., Pan, F., Lubrication performance of MoS_2 and SiO_2 nanoparticles as lubricant additives in magnesium alloy-steel contacts. *Tribol. Int.*, 93, 63–70, 2016.

7

Zinc Oxide Nanomaterials—Synthesis, Characterization, and Applications Focused on Lubricating Behavior

Monika Chauhan[1]*, Diwakar Chauhan[1], Ajay Kumar[2] and Arvind Kumar Jain[1]

[1]*Department of Chemistry, School of Basic Sciences, Galgotias University, Greater Noida, UP, India*
[2]*School of Biological and Life Sciences, Galgotias University, Greater Noida, UP, India*

Abstract

Zinc oxide (ZnO) is a white powder and broad bandgap semiconductor having an energy gap of 3.37 eV at room temperature. It is an excellent substance due to its high stability for chemical action, outstanding values for electrochemical coupling coefficient, broad range suitable for the absorption of radiations, and high response on exposure to solar radiation. ZnO are also used as nanolubricants as ZnO nanoparticles, which are prepared by various methods such as Fourier transform infrared spectrometry and X-ray Diffraction. They have been widely studied because of their tremendous applications in various areas like in electronics, pharmaceuticals, textile industry; and they have also been used significantly as lubricants in various industries. ZnO nanoparticles are made to act as nanoadditives, useful as nanolubricants and nanofluids. Synthesis of ZnO nanoparticles takes place by two broad routes: chemical and physical. Chemical methods include the hydrothermal method, sol–gel method, thermal decomposition, electrochemical depositions, microwave-assisted combustion method, ultrasound, combustion method, co-precipitation; and physical methods include laser ablation, chemical vapor deposition, anodization, and electrophoretic deposition. Formation of ZnO nanoparticles are confirmed by various tools X-ray diffraction (XRD), scanning electron microscopy (SEM), Fourier transform infrared spectrometry (FTIR), and atomic force microscopy (AFM).

**Corresponding author*: monika0283@gmail.com

Mohd Yusuf, Lalit Prasad and Shafat Ahmad Khan (eds.) Nanolubricants: Generation and Applications, (173–208) © 2024 Scrivener Publishing LLC

In this paper, we have described ZnO nanoparticle synthesis by physical and chemical methods. Characterization using XRD, SEM, AFM, optical transmission, and photoluminescence were studied. This paper also describes the nanolubricative nature of ZnO nanoparticles. ZnO nanoparticles are also synthesized through sol–gel roots as the main source of tribological evaluation of zinc oxide as a fundamental lubricant additive.

Keywords: Nanoparticles (NPs), zinc oxide (ZnO), synthesis, characterization, thin film

7.1 Introduction

Zinc oxide (ZnO) is an inorganic compound formed by two elements: zinc and oxygen. In the periodic table, it falls under the d and p block having groups 12 and 16, respectively. It arises naturally as the rock crystal zincite; utmost, zinc oxide formed unnaturally. It is a white-colored powder, insoluble in water, with wide applications. ZnO is an excellent substance due to its numerous properties, which are physical and chemical. Its stability for chemical action is very high with outstanding values for electrochemical coupling coefficient. It is also broad range, suitable for absorption of radiations, and it has a high response on exposure to solar radiations [1–11]. Zinc oxide is used as an additive for a range of products and consumer goods, including latexes, plastics, porcelain, glass, cement, paints, ointments, batteries and lubricants [12]. Zinc oxide has various lubrication additives and can be processed using various approaches. It plays imperative roles in the mechanism of engine system and fuel economy; it is also very effective in regulating lifecycles of various lubricants.

There are various manners in which zinc oxide plays an effective role in improving the characteristics of lubricants. Various nanomaterials and nanoscale-based materials act as nanoadditives for the betterment of solid and liquid lubricants such as ZnO, CuO, diamond, TiO_2, and Al_2O_3. Zinc oxide may also act as a nanolubricant and possess excellent anti-oxidative and anti-wear properties [13]. At nanoscale, zinc oxide (ZnO) particles with diameters less than 100 nm consist of a large surface-to-volume ratio ranging from 0.063 cm^2 to 0.188 cm^2 [14] comparative to their size, and shows high catalytic activity. Zinc oxide is a broad bandgap semiconductor material having an energy gap with a value of 3.37 eV at room temperature [15]. The structure of the ZnO occurs in chiefly two phases, namely, zinc-blende (cubic) and wurtzite (hexagonal). A commonly known structure is wurtzite, which is stable at pervasive circumstances [16]. Figure 7.1 shows

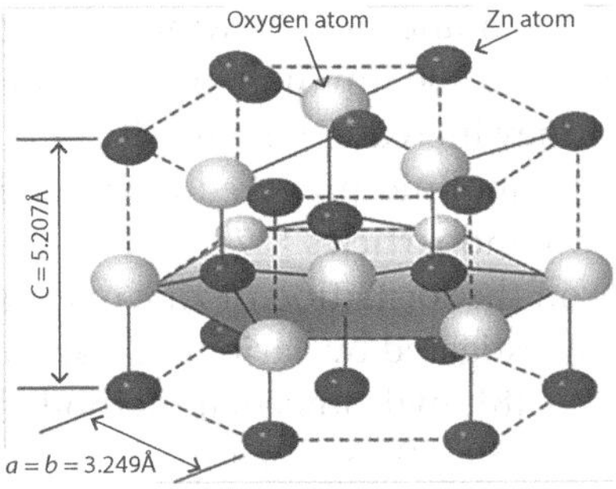

Figure 7.1 Crystal structure of ZnO hexagonal wurtzite.

the hexagonal wurtzite (B4) crystal structure form of ZnO, which consists of two lattice parameters, a and c. In this structure at the tetrahedron corners, every negatively charged species, i.e., anion, is enclosed by four positively charged species that are cations, which constitute the tetrahedron coordination [17].

ZnO piezo- and pyroelectric properties can be widely utilized as materials in various applications such as in the form of sensor material, energy converter, energy generation field, and as a photo-catalyst for the production of hydrogen in water splitting [18–21]. Zinc oxide is a material of interest for biomedicine and pro-ecological systems because of its low toxicity, biocompatibility, and biodegradability. It's an important material in the ceramics sector because of its hardness and stiffness [22–24]. Because of its rigor and constant high piezoelectric charge, it is a potentially important material for use in electronics, optoelectronics, and laser technology.

There are various tribological properties associated with ZnO nanoparticles, which are modified by oleic acid (OA), that act as surface capping agents for lubrication. Modified ZnO nanoparticles are used very effectively in reducing friction and increasing anti-wear abilities [13]. These OA-modified ZnO nanoparticles are widely prepared and useful in steel industries. These are prepared by various methods like X-ray diffraction and Fourier transform infrared spectrometry. ZnO nanoparticles synthesized through sol–gel roots are the main source of tribological evaluation of zinc oxide as fundamental lubricant additives [13, 25]. Various research have confirmed that the addition of certain nanoadditives such as dichalcogenides to lubricants are responsible for inflating the tribological characteristics. A major purpose of nanolubricants is to prevail over friction and anti-wear demerits [26]. Nanofluids are one of the most stable suspensors, categorized into Newtonian and non-Newtonian materials with the general particle size of less than 100 nm in diameter. A major characteristic of such nanofluids is that they increase the viscosity of lubricants. Turbine oil is one such example of nanofluid prepared commercially as a nanolubricant by dispersing ZnO nanoparticles [27].

Countless methods on the formation of zinc oxide structures in nanodimensions have been investigated by various researchers in literature such as laser ablation [28, 42–48], hydrothermal methods [29, 56, 57], electrochemical depositions [30, 59, 60], sol–gel method [31, 63–65, 67], chemical vapor deposition [32, 50, 51], thermal decomposition [33, 69, 70], combustion method [34, 35, 72], ultrasound [36, 73], microwave-assisted combustion method [37, 38], anodization [39], co-precipitation [40, 75], and electrophoretic deposition [41].

7.2 Preparations

The synthesis of ZnO nanoparticles takes place by two broad routes, i.e., chemical and physical. Chemical methods include the hydrothermal method, sol–gel method, thermal decomposition, electrochemical depositions, microwave-assisted combustion method, ultrasound, combustion method, and co-precipitation. Physical methods include laser ablation, chemical vapor deposition, anodization, and electrophoretic deposition.

7.2.1 Synthesis of ZnO by Pulsed Laser Ablation Technique

Pulsed laser deposition (PLD) is also known as the physical vapor deposition (PVD) method. In this method, a pulsed laser beam with high power focuses on a material taken as a target for the deposition of thin film inside a high vacuum chamber. The deposition of the thin film takes place on a substrate like a silicon wafer due to the vaporization of material from the target (in a plasma plume) as shown in Figure 7.2.

This process can also take place in the environment of O_2 gas, which is commonly used during the deposition of oxides to oxygenate the films to be deposited or it can be performed in ultra-high vacuum [41]. Several researchers [28, 42–48] reported the synthesis procedure of zinc oxide particles in nanodimensions using a pulsed laser ablation technique. In this technique for the dissociation of zinc targets, different researchers used various solvents such as double-distilled water [42, 46], surfactant, i.e., sodium dodecyl sulfate [44, 45, 48], NaOH [43], and starch [47]. The zinc

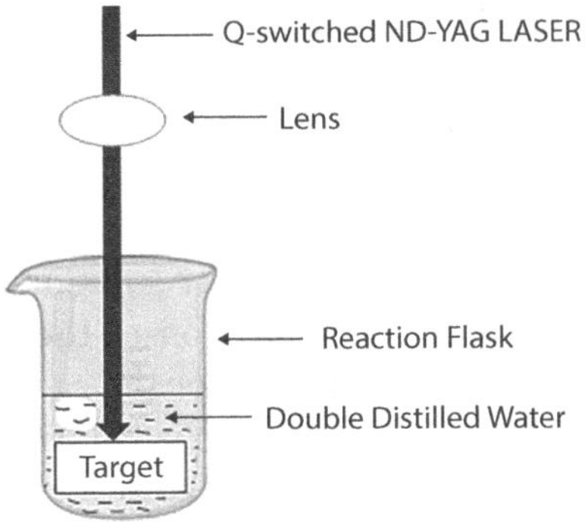

Figure 7.2 Pulsed laser ablation process.

target is placed at the lowermost part of the glass container ranging from 1 to 25 ml of DDW/surfactant/NaOH and starch as a solvent. For ablation, process Q-switched Nd-YAG LASER (Neodymium-doped yttrium aluminum garnet laser) was used. It works at a particular 1064 and 532 nm wavelength with a range of 6–15 Hz pulsed repetition rate and 7–15 ns pulse duration at an energy of a wide range 70–3209 mJ); the beam of the laser is directed towards the target substrate having a diameter ranging from 0.5 to 2 mm. For the ablation process, a lens having a focal length of 85–250 mm was used for laser focusing purposes by various researchers.

7.2.2 Synthesis of ZnO by Chemical Vapor Deposition Method

Chemical vapor deposition (CVD) [49] is a process in which desired films of volatile precursors are deposited on the surface of a substrate by the decomposition of chemicals shown in Figure 7.3.

Several investigators [32, 50, 51] proposed the preparation of ZnO nanopowder/nanoparticles/nanorods/quantum dots. For ZnO nanopowder, the starting material was zinc nitrate ($Zn(NO_3)_2.6H_2O$). The 0.1 molar

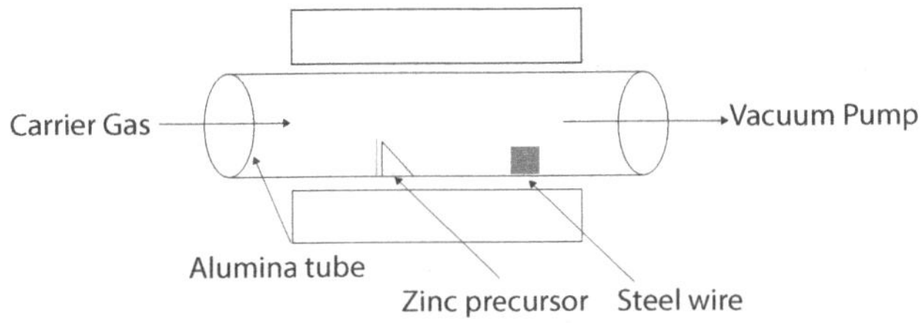

Figure 7.3 Chemical vapor deposition.

solution of a metal nitrate in aqueous medium passed into the reactor with the help of liquid atomization. The small-size drops, having the reactant material, were passed into the reactor and accumulated on a specific target for deposition. The ultrasonic nebulizer, having a frequency of atomization with a value of 1.7 MHz, was utilized to atomize a chemical solution for the flow of the small drops. Nitrates precursor solution was transferred into the vessel from the inlet side. The vibration of the transducer generated the aerosol. Lang expression was utilized for the calculation of the average diameter of small hazy drops. The equation is $Dd = 0.34\ (8\pi\ \gamma/\rho f^2)^{1/3}$ where Dd is the diameter of the small-size drop, c is the surface tension of the solution with a value of 72.9×10^{-3} N/m, q is the density of the solution with a value of 1,030.3 kg/m^3, and f is the employed ultrasonic frequency with the value of 1.7 MHz. The above parameters were utilized to calculate the diameter of the hazy droplets and as a result, the value obtained for droplets was at 2.8 μm. They produced small drops of hazy liquids, and were then transferred into the small compartment with the help of an air carrier gas, having a flow rate value of 1 dm^3/min, and then heated in the absence of oxygen at a predefined temperature value of 400°C. As a consequence, ZnO powder with the size range of 50–100 nm was obtained [32]. A. Forleo *et al.* investigated that colloidal ZnO nanocrystals were prepared using tetramethylammonium hydroxide pentahydrate $(CH_3)_4NOH{\cdot}5H_2O$ (TMAH, 2.5 millimole) as a starting material, which was incorporated into a 100% ethanol (5 ml), and then it was added drop-by-drop in a 0.1-M zinc acetate solution in 15 ml of dimethyl sulfoxide (DMSO), under vigorous stirring at room temperature, and finally, the nanocrystal of zinc oxide was washed using a non-aqueous solvent like ethyl acetate [50]. Siti Nor Qurratu Aini Abd Aziz *et al.* discussed ZnO nanorods' growth on the substrates of stainless-steel wire (ϕ:0.05mm) applying the CVD method. The wire was dipped in acetone, and then ultrasonically cleaned using acetone and deionized water. Pure Zn powder of 99.99% was used as a starting material. For the growth of ZnO NRs various masses of zinc powder at 0.3 g, 0.5 g, 1.0 g, 1.5 g, and 2.0 g, were used, respectively. Figure 7.3 illustrates the alumina boat, which was kept at the middle portion of the horizontal tube furnace. At the downstream portion of the furnace, the stainless steel wire was kept. The temperature and pressure value inside the furnace was kept at 650°C and 1 mbar, respectively, during the synthesis process. Oxygen gas in a trace amount (2–3 sccm) was supplied to the furnace for developing ZnO NRs. After 30 min, the oxygen supply was stopped to decrease the temperature value inside the furnace to room temperature [51].

7.2.3 Synthesis of ZnO by Anodization Method

The anodization method is a type of electrochemical method, which can be utilized for modifying the thickness of the oxide layer onto the metal surface. As the treated part of a metal acts as an anode in the electrolytic cell thus the method is also termed as anodizing as shown in Figure 7.4.

The anodization process improves resistance against corrosion and gives excellent adhesiveness for primers used in paint than those possessed by bare metal. Anodic thin films can also be utilized for various other purposes like in various effects in cosmetic products, either by the use of thick coatings having holes for the absorption of dyes or with thin coatings having transparency for the addition of interference effects to light in the reflected form [52]. Shetty *et al.* proposed the modest and consistent method of preparation of ZnO NPs by the anodization process.

In this process, the ZnO permeable structure was prepared at room temperature using a sheet of Zn mixed with distilled water as an electrolyte, with a pH of 5.8. In this process, graphite was used in the form of a counter electrode. Finally, the sheet of Zn is anodized with various voltages in the range of 1 to 9 volts. The time used for deposition was varying in the range of 6 to 12 h. The time used in the reaction for applied voltage having a value of 3 V is diversified in the range of 6, 8, 10, and 12 h [39].

Application of voltage for an extended time leads to the dissolution of zinc oxide, and therefore the value of current density increases and reaches nearly up to saturation. Over time, the thickness of the porous film, which is measured using the Dektak profilometer, decreases. For 12-h deposition time, the calculated thickness value is 7.8 μA, with an average etching rate of 0.65 μm/h. It was found that the ZnO film morphology depends enormously on the current density [53].

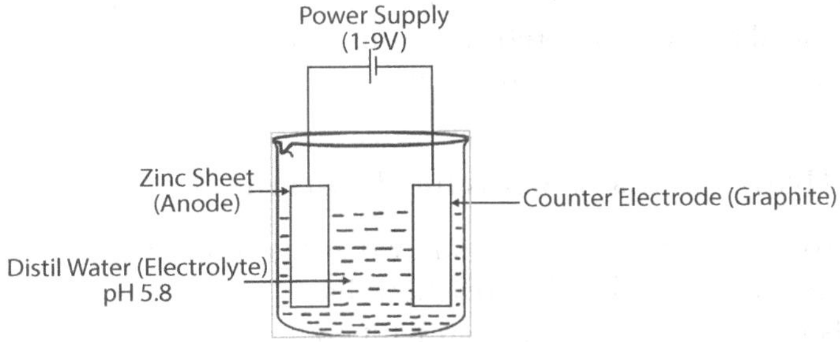

Figure 7.4 Anodization process.

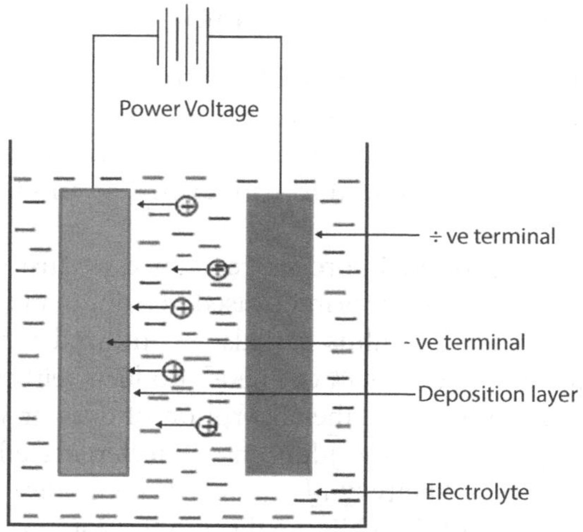

Figure 7.5 Electrophoretic decomposition technique.

7.2.4 Synthesis of ZnO by Electrophoretic Deposition Process

The electrophoretic deposition (EPD) process is a colloidal process where a thick or thin coating/film of charged particles is deposited onto a conductive substrate under the influence of the electric field. A large number of colloidal particles of the composites, metals, polymers, and ceramics can be applied for the EPD process. Applied potential and deposition time can also be adjusted to gain easy control over the thickness of coating/film and its morphology as shown in Figure 7.5 [54].

Alejandro Vazquez *et al.* demonstrated the growth mechanism of 1D zinc sulfide structures in the nanodimension range by applying the electrophoretic deposition method. Zinc sulfide NPs having a mean diameter of 20 nm were synthesized in the microwave-assisted method with the help of sodium citrate, which acted as a stabilizer. After that, aqueous dispersion was deposited by electrophoretic methods in which aluminum plates were used as electrodes under 600 mV [39].

7.2.5 Hydrothermal Process for the Synthesis of ZnO

The hydrothermal process is also known as hydrothermal procedure and can be described as a chemical reaction that is used to synthesize various materials in aqueous media at high temperature and pressure performed in a sealed pressure vessel called an autoclave [55]. Various reports investigated show the formation of zinc oxide particles in the nanodimension

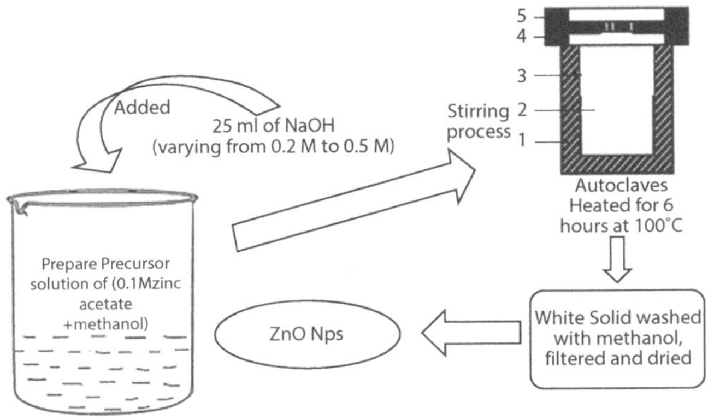

Figure 7.6 Hydrothermal method.

range by applying the hydrothermal process [56, 57] in which zinc acetate and methanol are used as starting materials. In this method, a 0.1-M zinc acetate solution was prepared in a 50-ml methanol with different concentrations (varying from 0.2 M to 0.5 M) [56]; this was then added to 25 ml of NaOH under study. This solution was completely poured into a sealed autoclave made up of stainless steel with Teflon lining and heated for 6 h at 100°C under autogenous pressure. At room temperature, this solution was placed to cool down. The resultant products obtained in the form of white solid were washed using methyl alcohol, filtered, and further air dried in a laboratory-type oven at 60°C [57] as shown in Figure 7.6.

Yong-hong Ni *et al.* reported that a $ZnCl_2$ solution with 0.001 mol concentration and potassium hydroxide (KOH) solution with a 0.002 mol concentration were prepared in double-distilled water, and these solutions were used to form zinc oxide nanorods. In this solution, 0.0005 mol of cetyltrimethylammonium bromide (CTAB) was added with stirring. This solution was poured into the sealed Teflon-lined stainless steel autoclave with 40-ml capacity filled with distilled water that occupied up to 80% volume. This reaction proceeded oat a temperature of 120°C for 5 h. After cooling the solution, white precipitates thus obtained were washed with distilled water and ethanol, and was finally dried for 5 h at 50°C [29].

7.2.6 Synthesis of ZnO by Electrochemical Deposition Method

The electrochemical deposition, also known as electroplating, is a process that forms an *in situ* metallic, oxide, or salt coating above a conductive material surface by using a solution consisting of the desired metal ion or

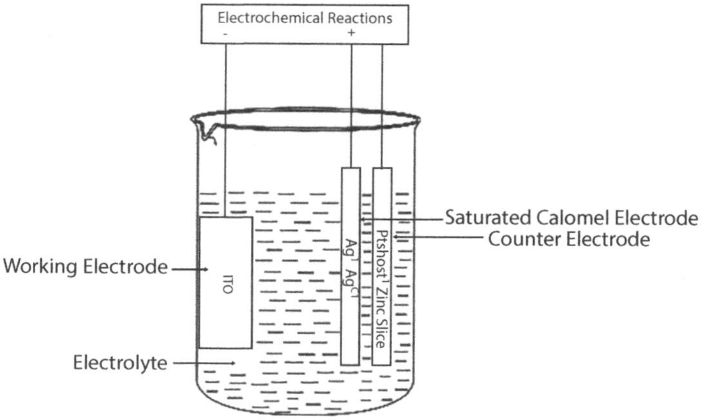

Figure 7.7 Electrochemical deposition method.

its chemical complex [58]. ZnO NPs are prepared by the electrochemical deposition process using various types of working substrates like silicon coated with gold, indium oxide doped with tin oxide (ITO), and a strip of copper metal [30, 59, 60] in which ITO is the most commonly used substrate. The ITO conductive material surface has a sheet resistance of (10 Ω/square), which is converted into small strips by cutting, having a width and length of 1.0 cm and 2.0 cm, respectively. This is processed by a three-electrode glass cell system, i.e., a working electrode, reference electrode, and a counter electrode, which is dipped in a water bath at a temperature of 70°C–80°C, shown in Figure 7.7.

Many investigators used ITO as a working electrode, Ag/AgCl/mercury/mercurous chloride electrode as a reference electrode, and Pt sheet/Zinc piece used in the form of a counter electrode. For the deposition process, zinc nitrate ($Zn(NO_3)_2.6H_2O$) with KCl as an electrolyte was used [59] or a solution of supporting electrolyte having $ZnCl_2$ (different molar ratio) in 0.1-M potassium chloride for depositing the film [60]. Final products thus obtained were rinsed and dried in a drying cabinet at 60°C.

7.2.7 Preparation of ZnO by Using the Sol–Gel Technique

The sol–gel technique is a wet-chemical process involving the formation of sol (inorganic colloidal solution) and gel (gelation of the sol in continuous liquid phase) to develop a 3D network structure [61]. This technique is very effective in modifying the surface of the substrate. One of the most important advantages of this technique (sol–gel) is obtaining high surface area and stable surfaces. The sol–gel technique has two main reactions: (1) precursor hydrolysis in both acidic and basic mediums and (2) hydrolyzed product's polycondensation. Through these reactions, metal nanoparticles (MNPs) are retained by polymeric networks [62]. Numerous reports show the utilization of the sol–gel method to synthesize the ZnO particles in nanodimensions and thin films. It was reported that ZnO powders were synthesized using an aqueous solution of zinc 2-ethyl hexanoate ($C_{16}H_{30}O_4Zn$) dissolved in 2-propanol, followed by the addition of tetramethylammonium (TMAH) aqueous solution [31]. In another process, a suitable proportion of triethanolamine is mixed into ethanol. After that, zinc acetate dihydrate (ZnAc) was added. To get the best ratio, the concentration of zinc acetate in sols was diversified to 0.25, 0.5, and 0.75 M, and the molar ratio of TEA to ZnAc was changed to 3:5, 6:5, and 9:5. For half an hour at 50°C–60°C, the solutions were heated and stirred to get ZnO NPs [63].

For the formation of a thin film of ZnO, 0.2-M zinc acetate 2-hydrate solution in colloidal form was prepared by dissolving in dry ethanol and refluxed for 3 h at temperature of 80°C, and 2-methoxy ethanol and C_2H_5OH were mixed in the resulting precursor. Then, for stabilization, the solution used as a precursor was kept overnight and utilized for the deposition of a thin film with a spray-coating method, shown in Figure 7.8, at the prepared film having properties of ZnO. On increasing or decreasing the sintering temperature of the film ranging from 400°C–600°C, the thickness of the film varies from 2.37 to 2.83 μm. The obtained film thickness ranged from 2.37–2.83 μm. The measured density of films ranged from 2.24–2.69 g cm^{-3}, which is ≈ 40%–48% of the theoretical density (5.60 g cm^{-3}) of ZnO. This indicates that the films are porous and may provide a wider electrolyte contact area when used in photoelectrochemical cell (PEC) cells [64]. Mn-doped ZnO nanoparticles were prepared by adding 2.0 g of zinc acetate dehydrate in 14 ml of CH_3OH and stirred for t2 h followed by the addition of 0.083 g of $MnCl_2{\cdot}4H_2O$ by a continuous process of stirring to completely dissolve the materials taken as precursors. Further, the obtained solution was finally dried in high-pressure vessels under supercritical conditions of ethanol, at the heating rate of 45°C/h to produce aerogel powder [65]. ZnO: Mn nanoparticles were obtained with variable fine particle sizes, and with the increase in the concentration of doping, an increase in particle size was observed. The size of the majority of ZnO: Mn particles in this powder was between 17 and 34 nm [66]. Riyadh M. Alwan *et al.* demonstrated the preparation of ZnO nanoparticles' hydrolysis of 12.6 g of zinc acetate dihydrate with 400 ml of double-distilled water by continuously stirring and heating this solution at 50°C, then with 600 ml of absolute alcohol that was constantly stirred, added slowly. After this, a clear solution of 6 ml of H_2O_2 (47%) was added drop by drop and mixed using a magnetic stirrer.

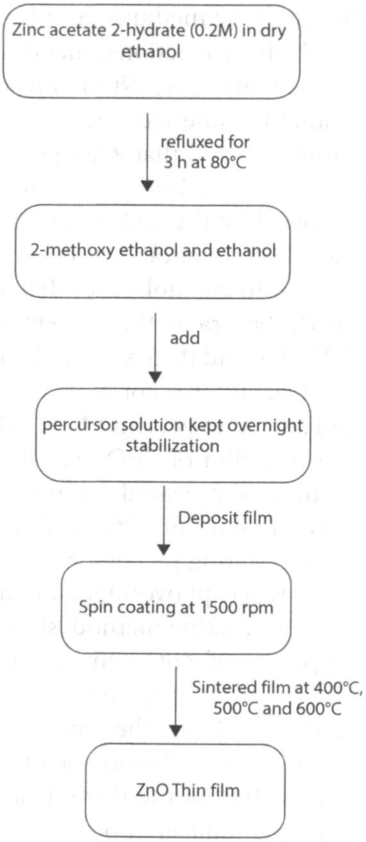

Figure 7.8 Sol–gel method for the ZnO thin film.

White zinc oxide nanoparticles were obtained by incubating this solution for 24 h and followed by drying at 80°C for several hours [67].

7.2.8 Synthesis of ZnO by Thermolysis Method

Thermolysis, which is known as *thermal decomposition,* is a type of chemical reaction, which takes place by *the breakdown* of chemical species by the action of *heat*. The temperature at which the substance is heated and decomposed by chemical action is known as *decomposition* temperature. The reaction is referred to as an endothermic reaction as the compound undergoing *decomposition heat* is required to break chemical bonds [68]. Various researchers [33, 69, 70] used zinc nitrates/zinc acetate as a precursor for the preparations of ZnO nanoparticles. Ruey-Chi-Wang *et al.*

first reported that zinc oxide particles in nanodimensions, nanowires, and nanowalls can be prepared on the surface of a silica substrate through the thermal breakdown of $Zn(CH_3CO_2)_2$ by an altered CVD method at a temperature between 200°C and 250°C [33]. Davood Raoufi described the synthesis of ZnO nanoparticles by using the solution of Zn $(NO_3)_2$ with ammonium carbonate $(NH_4)_2CO_3$, having a ratio of 1:1.5 under stirring. Formed white precipitates were washed using H_2O and C_2H_5OH, filtered, and further dehydrated at 100°C for 6 h to get zinc oxide, which is used as a precursor. Finally, the precursor heated in the muffle furnace at a temperature ranging from 250°C to 550°C for 4 h obtained zinc oxide particles in the nanodimensions range as shown in Figure 7.9 [69]. ZnO nanopowder is also prepared using zinc carbonate ($[ZnCO_3]_2$. $[Zn(OH)_2]_3$) as a precursor by heating at 600°C for 1 h in a furnace [70].

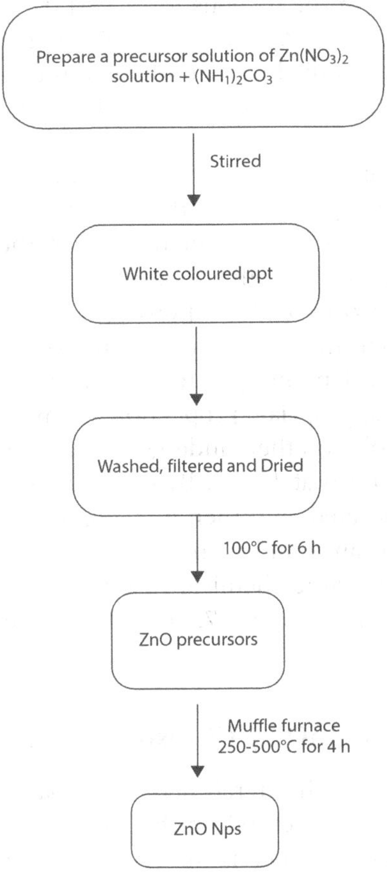

Figure 7.9 Thermal decomposition method.

7.2.9 Synthesis of ZnO by Combustion Method

Volume combustion synthesis (VCS) is a chemical method in which the reaction materials are uniformly heated in controlled conditions until a reaction ignites throughout the whole volume of the reacting chamber. The VCS mode is mainly applied for weak exothermic reactions in which preheating is required before ignition [71]. For the preparation of ZnO nanoparticles, $Zn_{1-x}Fe_xO$ (where x = 0.00, 0.005, 0.01, 0.02, 0.03, 0.04) powders were prepared using stoichiometric composition of nitrates of metal as oxidizers and 2-aminoethanoic acid as a fuel. After that, the sample was placed in a dish made up of Pyrex glass, and then heated at a temperature of 400°C ± 10°C with the help of a muffle furnace. Finally, the combination becomes permeable, and huge powder is formed. The chemical reaction involved during the preparation method is like Zn $(NO_3)_2$ + Fe $(NO_3)_2$ → $Zn_{1-x}Fe_xO$ + products in the gaseous form [34]. K. Manjunath *et al.* proposed that the ratio of (1:1) Zn $(NO_3)_2$ and pigeon pea ground in the powder were dissolved in distilled water under continuous stirring process for about 10 min to obtain a homogenous solution, which is then heated at 120°C for 1 h to acquire a mass having very small bubbles. Then it is preheated at a temperature value of 300°C in a muffle furnace. After heating, observed flaming of the gel indicates that, within 3 to 4 min, nanocrystalline ZnO is obtained. To eliminate the impurities, the obtained product is calcinated for 2 h, and finally, milky-white ZnO nanoparticles are obtained [35]. Nanocrystalline zinc oxide in powder form was synthesized using a gel-combustion synthesis procedure using zinc nitrate tetrahydrate, $ZrOCl_2$, and citric acid monohydrate as reactants. In this method, the starting materials in appropriate proportion are mixed in double-distilled water. The aqueous solution then undergoes a uniform stirring process for the duration of about 2 h at 120°C. By continuous heated stirring, water was completely evaporated, and then it changes to gel. The gel augments into a foam-like substance and turns into fine powder by undergoing a self-propagating exothermic chemical reaction at a temperature value of 200°C, which is shown in Figure 7.10, that is heated for improving the ordered phase [72].

7.2.10 Synthesis of ZnO by Ultrasonic Method

A wide survey reported in the literature is the preparation of ZnO nanoparticles using the ultrasonic method. In this method, zinc acetate dihydrate was mixed with distilled water in three different molar concentrations (0.01, 0.1, and 1 M) used as a precursor. The precursor, which is used as

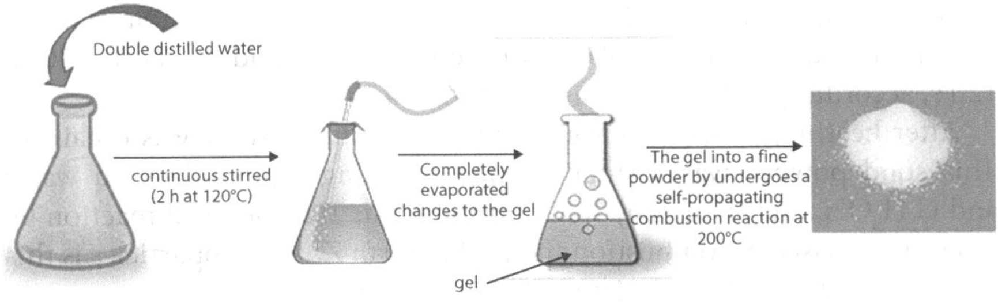

Figure 7.10 Synthesis of ZnO by combustion method.

a spray, was kept in a bath located at the higher part of the equipment of spray pyrolysis. Ultrasonic vibrations generated the spray of precursor solution, and the resultant drops were then passed into a high-temperature furnace tube kept at 900°C for 2 h [36]. H. R. Liu *et al.* also reported the preparation of zinc microphores by an ultrasonic hydrothermal process. In this method, an aqueous solution of zinc acetate ($Zn(Ac)_2$) and triethanolamine (TEA) was prepared, which was mixed with the help of a stirring process on a magnetic stirrer. After that, the solution was exposed to agitation by using ultrasonic waves with 200 W power having a 20-MHz frequency for a divergent time duration of 30, 60, 90, and 120 min, distinctly, and after it precipitates, washed by using distilled water and C_2H_5OH many times by applying centrifugation and was further dried in an oven under vacuum condition [73].

7.2.11 Microwave-Assisted Combustion Method to Synthesize Zinc Oxide

Various investigators proposed that zinc oxide can also be synthesized by applying a microwave-assisted combustion process as shown in Figure 7.11 [37, 38]. In this method, a mixture of zinc nitrate [$Zn(NO_3)_2.6H_2O$] and

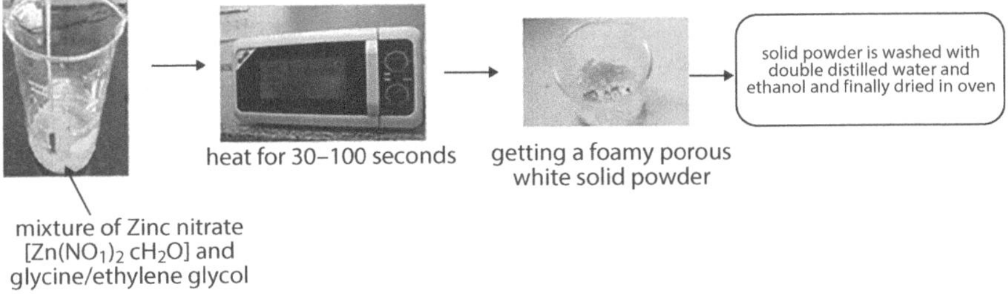

Figure 7.11 Microwave-assisted combustion method.

2-aminoethanoic acid/ethylene glycol with their respective molar masses were prepared, then it was placed in the microwave and heated for 30 to 120 s, according to their heat supply in the microwave.

After heating, a foamy, porous, and white solid powder was obtained. This solid powder was further washed by using double-distilled water and C_2H_5OH, and then dehydrated in the oven. The chemical reaction by microwave-assisted combustion method to form ZnO nanoparticles is this [38]: $Zn(NO_3)_2 \rightarrow ZnO_2 + 2NO_2$ (at a lower temperature)

$$2\,ZnO_2 \rightarrow 2ZnO + O_2.$$

7.2.12 Synthesis of ZnO by Co-Precipitation Method

Coprecipitation (CPT) or co-precipitation is the carrying down by precipitate of soluble substances under the conditions employed [74]. Various scientists [40, 75] investigated the preparation of ZnO nanoparticle by the co-precipitation procedure. Recently, G. Vijayaprasath *et al.* proposed that in this process, pure ZnO and ZnO doped with Gd were prepared followed by zinc nitrate [$Zn(NO_3)_2$], gadolinium nitrate[$Gd(NO_3)_2$], and sodium hydroxide(NaOH), and were used as starting materials, shown in Figure 7.12. In this method, in the solution of zinc nitrate,

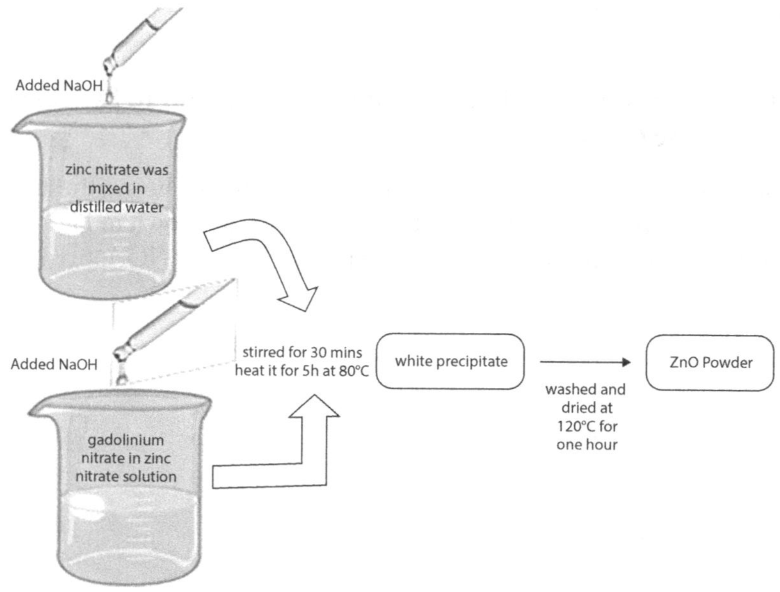

Figure 7.12 Synthesis of ZnO by co-precipitation method.

double-distilled water, and NaOH, the solution was added dropwise into it under stirring. The solution was mixed and stirred for 30 min at room temperature, and then was heated for 5 h at 80°C. A white precipitate was formed. Then, the precipitate was washed and dried at 120°C for 1 h. After that, the samples of ZnO in powder formed were heated and cooled at 300°C in the presence of oxygen for 2 h. For synthesizing $Zn_{1-x}Gd_xO$ (x=0.03, 0.06 and 0.09) particles in nanodimensions, a mixture of gadolinium nitrate solution was added into zinc nitrate, then a dropwise of a solution of sodium hydroxide was added, then a white precipitate formed, and further treatment was given similarly to ZnO. Again, the same method for the synthesis of undoped ZnO nanoparticles was applied for the preparation of ZnO nanoparticles doped with Gd. [40]. In another study, a similar method adopted to prepare pure ZnO nanoparticles was utilized to synthesize ZnO doped with Ni in nanodimensions. For synthesizing, pure Ni-doped ZnO, Mn-doped ZnO, and Mn co-doped ZnO/Ni were prepared. Sodium hydroxide (NaOH), zinc acetate dihydrate[$Zn(CH_3COO)_2.2H_2O$], manganese acetate dihydrate [$Mn(CH_3COO)_2.2H_2O$], and nickel acetate dihydrate [$Ni(CH_3COO)_2.2H_2O$] were used as starting materials. In this method, an NaOH solution and double-distilled water were added dropwise into the solution of zinc acetate. After a few times, the solution formed a white precipitate (ppt), which was stirred at room temperature for 30 min and then kept at 80°C for 5 h. Then, the precipitate was washed and dried at 120°C for 1 h. The prepared powder was annealed in the air for 2 h at 400°C. For synthesizing nanoparticles of Zn0.80Ni0.20O (Ni/ZnO), a mixture of nickel acetate in water was added to the solution of zinc acetate, then a drop-by-drop addition of aqueous NaOH solution was carried out to get a white ppt with a pale green color. Afterward, a similar method was adopted for the preparation of pure ZnO NPs for synthesizing Ni-doped ZnO NPs. Zn0.80Ni0.10Mn0.10O (Mn-Ni/ZnO) and Zn0.80Mn0.20O (Mn/ZnO) NPs followed the same principle adopted for the preparation of Ni-doped ZnO. Mn-doped ZnO and Mn co-doped ZnO/Ni NPs were obtained from a homogeneous mixture and a light brown and black-colored ppt was obtained, respectively [75].

7.2.13 Synthesis of ZnO by Green Synthesis Method

Green chemistry uses environmentally friendly materials and maintains compatibility with non-harmful chemicals for pharmaceutical and other biomedical applications. To synthesize various non-metallic NPs, a more energy-efficient and competitive biologic approach that involves many different types of microorganisms has been used in contrast to other chemical approaches. In biomedicine and related fields, the biocompatibility of bio-inspired NPs provides very important uses [76]. Biocompatible with NPs developed by chemical procedures is the coating of biomolecules on the surface of NPs [77–79]. Vidya C. proposed the green synthesis of ZnO nanoparticles with the help of a leaf extract from *Calotropis gigantea* plants. The extract was prepared with 50 g of leaves with 100 ml of double-distilled water in a 250-ml glass beaker, which was boiled for 3,600 s until the aqueous solution's color changed to light yellow, then it was cooled and filtered using filter paper. For the preparation of nanoparticles, 50 ml of the extract was boiled at 60°C–80°C with the help of a stirrer-heater. As the temperature reached 60°C, zinc nitrate (5 g) was added to the solution. The mixture was boiled until the color changed to a light-yellow solution, and then to a deep yellow paste. This paste is put in a ceramic crucible and heated for at least 2 h in a furnace at 400°C to obtain a light yellow-colored powder of ZnO nanoparticles as shown in Figure 7.13 [80].

In a study, Fakhari adopted a co-precipitation procedure to synthesize ZnO nanoparticles [81]. From the family *Lauraceae*, *L. nobilis L.* fresh leaves were washed and then dried. The extract was obtained by mixing powder leaves of 20 g with 200 ml of deionized water with continuous stirring on a hotplate for 10 min at a temperature value of 80°C. For the removal of plant remaining parts and impurities from the mixture, the filtration and centrifugation processes were used, and after that, the mixture

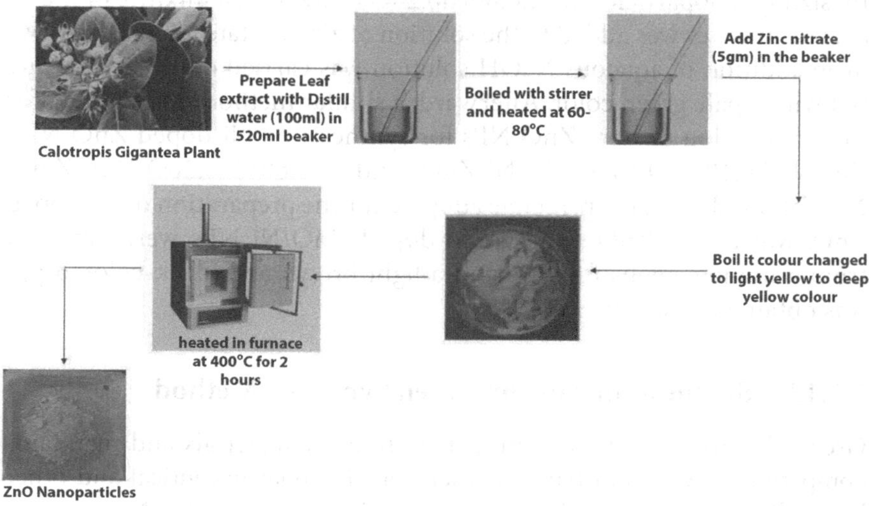

Figure 7.13 Preparation of ZnO powder with *Calotropis Gigantea* plant.

was stored for further utilization. To synthesize zinc oxide nanoparticles from zinc acetate and plant extract, a mixture of 1 ml of aqueous extract of *L. nobilis* leaves and a 25 ml of 0.02-M $Zn(OAc)_2.2H_2O$ solution was used. A magnetic stirrer was utilized to continuously stir the mixture at room temperature for 2 h. Further, 0.02-M NaOH was added dropwise to maintain a pH value of 12. To obtain a light yellow-colored solid product, the mixture was stirred for one extra hour. Purification of the obtained precipitate was done with deionized water using multiple redispersion methods and was further centrifuged. A white powdered material was obtained as a final product, and it was kept overnight in an electric oven for drying at 60°C. Another green synthesis study of zinc oxide nanoparticle by zinc nitrate was performed using a mixture of 4 ml of aqueous extract of *L. nobilis* leaves with 25 ml of 0.05-M Zn $(NO_3)_2.6H_2O$ solution, and then the reaction was further continued [82].

7.3 Characterization

ZnO nanoparticles have been characterized by using various instrumental techniques. The properties of ZnO nanoparticles have been [83–89] characterized using XRD, SEM AFM, optical transmittance, and photoluminescence. Figure 7.14 shows the X-ray diffraction (XRD) spectrum

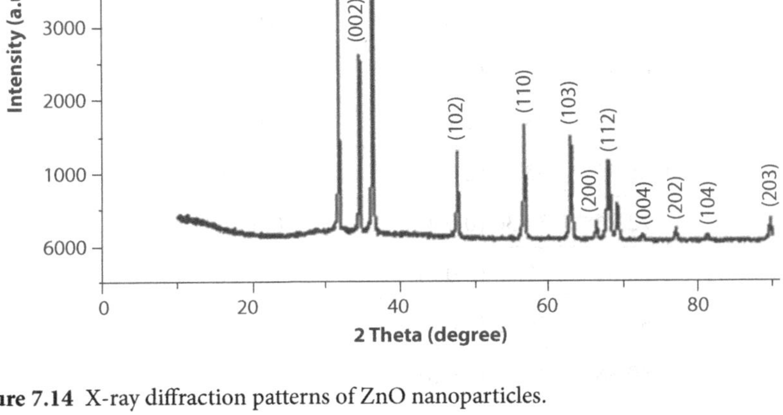

Figure 7.14 X-ray diffraction patterns of ZnO nanoparticles.

of the biocompatible nanoparticles of zinc oxide synthesized through the green process. The main peaks corresponding to Bragg reflections were found with 2θ values of 31.74°, 34.38°, 36.21°, 47.49°, 56.54°, 62.86°, 66.43°, 67.89°, 68.99°, 72.76°, 76.94°, 81.32°, and 89.56°. The Bragg reflections' characteristic locations were indexed to (1 0 0), (0 0 2), (1 0 1), (1 0 2), (1 1 0), (1 0 3), (2 0 0), (1 1 2), (2 0 1), (0 0 4), (2 0 2), (1 0 4), and (2 0 3) planes of ZnO hexagonal phase structures, respectively, standard JCPDS card 36-1451 and this confirmed the presence of nanoparticles of zinc oxide. No other impurities are shown in the XRD pattern. Table 7.1 shows the crystallite sizes and lattice strain values for the nanoparticles of ZnO, which is calculated by Debye Scherrer's formula that can be denoted as:

$$D= 0.94\lambda / \beta \cos \theta$$

where D indicates the average crystallite size, β indicates the radians line broadening, θ indicates the Bragg angle, and λ indicates the X-ray wavelength.

Table 7.1 Crystallite size and lattice strain values of ZnO nanoparticles derived using the XRD method.

2θ (°)	Crystallite size (nm)	Lattice strain
31.74	28.77 (1 0 0)	0.0046
34.38	28.96 (0 0 2)	0.0042
36.21	29.11 (1 0 1)	0.0040
47.49	30.23 (1 0 2)	0.0030
56.54	31.41 (1 1 0)	0.0024
62.86	32.42 (1 0 3)	0.0021
66.43	33.07 (2 0 0)	0.0020
67.89	33.35 (1 1 2)	0.0019
68.99	33.57 (2 0 1)	0.0019
72.76	34.36 (0 0 4)	0.0018
76.84	35.33 (2 0 2)	0.0016
81.32	36.47 (1 0 4)	0.0015
89.56	38.97 (2 0 3)	0.0013

Figure 7.15 SEM images of ZnO thin films.

The average crystallite size observed from the intense plane (1 0 1) was measured as 29.11 nm. X-ray diffraction patterns determined through this analysis are similar to the X-ray diffraction determined for the nanoparticles of ZnO with biological synthesis [84]. SEM was used to get information about the surface morphology and composition of the material [85].

In the leaf extract of *Ocimum tenuiflorum*, hexagonal particles with a diameter ranging from 11 to 25 nm were found by researchers [86]. Figure 7.15 displays the SEM images of synthesized ZnO nanoparticles.

The SEM images display aglomerations of individual particles of the zinc oxide. A close inspection of the agglomerated lump indicates the presence of many aggregates of nanoparticles. In Figure 7.15(a), particles appear to be agglomerated and some individual crystals are visible. In Figure 7.15(b), some particles appear to be elongated, hexagonal, and rod-shaped.

A widely used technique in the characterization of ZnO nanoparticles is atomic force microscopy (AFM), which is also utilized to identify the surface morphology. Figure 7.16 represents the AFM surface framework pictures of the ZnO thin films having divergent aging time, i.e., 0 h, 72 h, 168 h, and 312 h. AFM pictures indicated that there are no cracks nor tiny holes there. All the films show smooth surface morphology. AFM analysis identifies, for all samples, the root-mean-square of average surface irregularity, which determined a scanning area of $5 \times 5\ \mu m^2$. In ZnO thin films, the surface irregularity factors measured were 10.67, 5.07, 5.72, and 7.33 nm that correspond to the divergent aging time, i.e., 0, 72, 168, and 312 h, respectively. The data obtained through this analysis showed that in ZnO films, on increasing the sol aging time, irregularity in the surface is decreased [87]. The optical transmission spectra of zinc oxide thin films were verified at 300–700 nm wavelength, having divergent sol aging time, as depicted in Figure 7.16. The deposited ZnO films have a highly smooth and clear appearance that shows an high transmission above 70%–80% in the range of visible region, which moderately gets better with increasing time, as this indicates that with an increasing optical transmittance, surface irregularities of films decrease, which are identified by AFM pictures. Figure 7.17 represents that observing ZnO films in UV range, considering the different sol aging time with a duration of 0 to 168 h, the significant

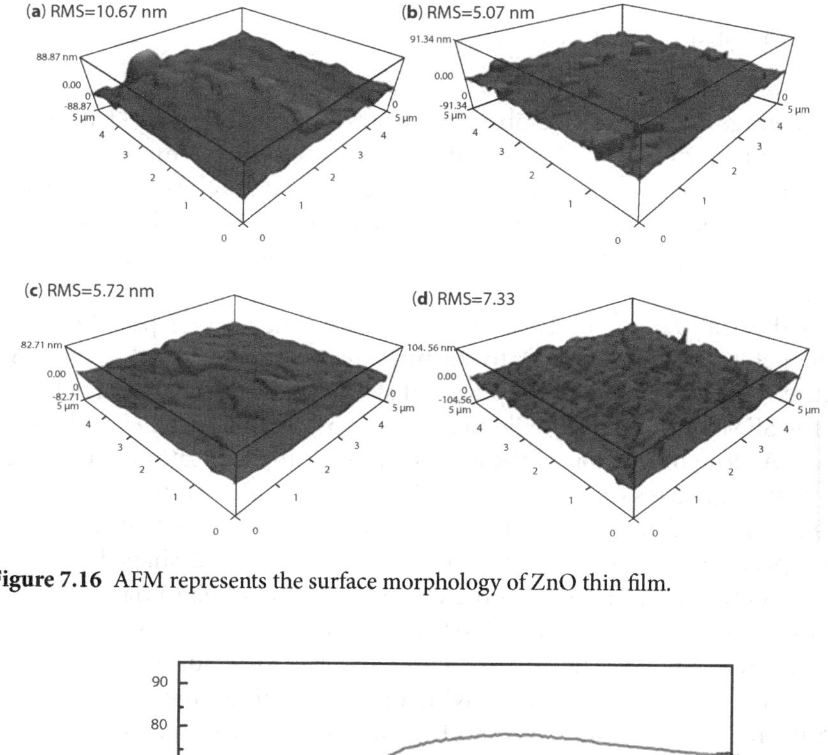

Figure 7.16 AFM represents the surface morphology of ZnO thin film.

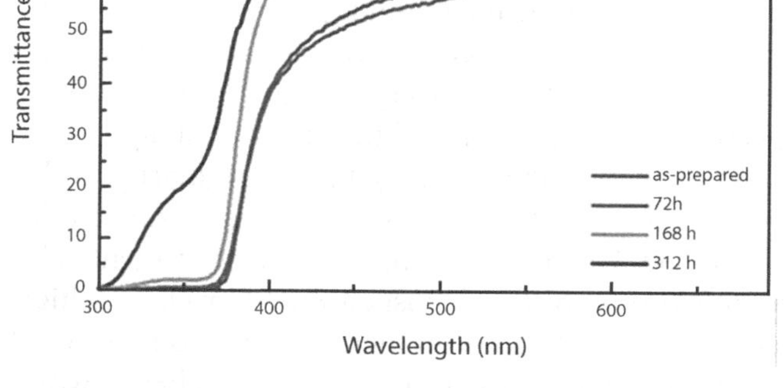

Figure 7.17 Optical transmission spectra of ZnO thin films.

absorption property of the films was not affected. The observations of all the samples also reveal that important absorption edges were observed at a value of ~370 nm that relate to the ZnO optical bandgap transition.

Like ZnO, i.e., the direct transition materials, the absorption coefficient (α) can be expressed by utilizing the Tauc model in the region of high absorption [88] $(\alpha h\nu)^2 = A\,(h\nu - E_g)$ where α denotes the coefficient of optical absorption, h denotes the Planck's constant and ν denotes the frequency, Eg denotes the optical band gap, and A denotes the direct transition constant. By assuming that the absorption coefficient α~- Ln(T) corresponds to the direct bandgap for Wurtzite structure, a plot between $(\alpha h\nu)^2$ and the photon energy hν represents a sharp absorption edge for the good quality thin films by a linear fit [89]. Figure 7.18 displays the plots between $(\alpha h\nu)^2$ and hν for the thin films of ZnO prepared using sol for various time intervals. The extrapolation of the $(\alpha h\nu)^2$, plot verses hν to the energy axis calculate the optical band gap (Eg). The deviation of bandgap energy for the sol aging time is shown in the inset of Figure 7.18, which represents a rise in the value of optical energy gap from 3.19 to 3.25 eV takes place considering the sol aging time in between the duration of 0 and 312 h, respectively [87]. Changing the dimensions of semiconducting materials to a nanoscale takes place in the physical properties of semiconducting materials, which is referred to as "quantum size effects," e.g., by quantum internment, the bandgap energy of ZnO increases and is clearly shown from photoluminescence [90, 91]. Figure 7.19 shows the photoluminescence spectrum of zinc oxide in nanopowder form at room

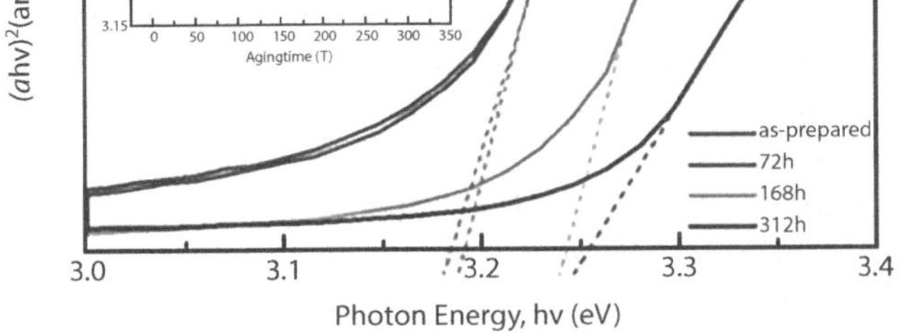

Figure 7.18 Plot of (αhν)2 vs. photon energy (hν) of ZnO film with different sol aging time.

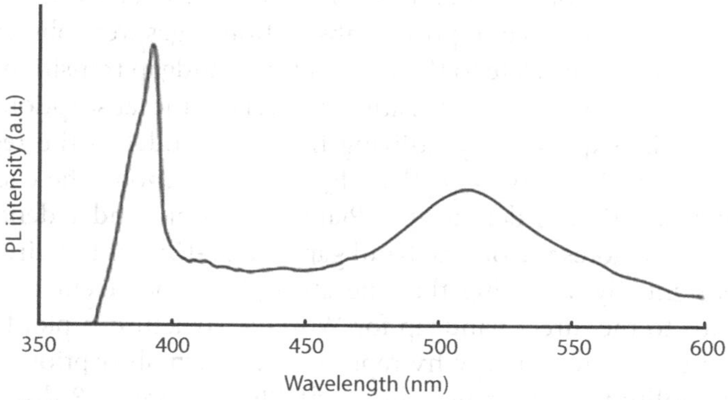

Figure 7.19 Photoluminescence spectrum of ZnO.

temperature, with an aggravation wavelength value of 320 nm. The spectrum shows two peaks of emission: one nearly 392 nm (Ultraviolet region) and the other at around 520 nm, which are commensurate to the excitonic emission near the bandgap [92] and represent the presence of singly ionized oxygen vacancies, respectively [93]. The radiative recombination of a photogenerated hole having an electron residing in the oxygen vacancy is mainly responsible for the emission [92, 94]. In addition to that, the spectrum also shows the small size of particles in the nanodimension in powder form as the luminescence peak full-width at half-maximum (FWHM) with its value only in a few nanometers [95].

ZnO nanoparticles have also been characterized as nanolubricants in many industries such as the automobile industry. It was well recognized in the beginning of the 19th century, only that automobile engines consume around 30% of fuel energy as friction power waste [96]. Hence, the ZnO nanoparticles are made distinctive by their tribological characteristics used as nanoadditives, which result in controlling wear and friction trends. This is actually a result of a chemical process in which a tribofilm layer is being formed on the sliding surfaces in order to separate the two surfaces to prevent friction [96, 97]. Various nanoparticles, which are used as lubricants, are found to have additional characteristics, i.e., when nanoparticles are optimally used as lubricants, they possess high viscosity, have low frictional coefficient, have low wear scar diameter abilities, and many more [97]. Lubricating oil is the most crucial finding of nanolubricant itself, which is the most supportive material in modern industry.

7.4 Applications

ZnO is broadly used in various fields. It plays a key role in applications like the electronic industry, pharmacy and cosmetics industry, rubber industry, textile industry, photocatalyst, and sensors, as shown in Figure 7.20, because it possesses different chemical and physical properties [98].

Opto-electronic industry. ZnO possesses good exciton-binding energy, having 60 meV value, therefore it can be used as a possible interesting material for various light-emitting devices in the f ultraviolet (UV) region. As ZnO has a direct and broad bandgap, having a value of 3.37 eV, it has wide applications and is used as a material in sensors, in solar cells, in photodetectors, and for making light-emitting diodes [99–101].

Pharmacy and cosmetic industry. Zinc oxide possesses good antifungal and antibacterial activity, so it is extensively utilized in the manufacture of numerous basic materials, which can be utilized in the form of medicine, like germ-killing substances and for curing various skin diseases [102]. Because the UV region is suitable for ZnO bandgap, radiations in the UV region can therefore be easily absorbed by ZnO nanoparticles, which can be widely used as a material in the production of sun-protective creams [99, 103].

Concrete and rubber industry. Zinc oxide can also be utilized as a potential material for the manufacture of concrete because zinc oxide reduced the processing time of concrete and enhance its water resistance property. Additive zinc oxide is used in the rubber industry. It acts as a catalyst for the vulcanization of sulfur by boosting the efficiency of the cross-linking system. The mechanical and thermal properties of silicone rubber can be improved by incorporating zinc oxide filler in

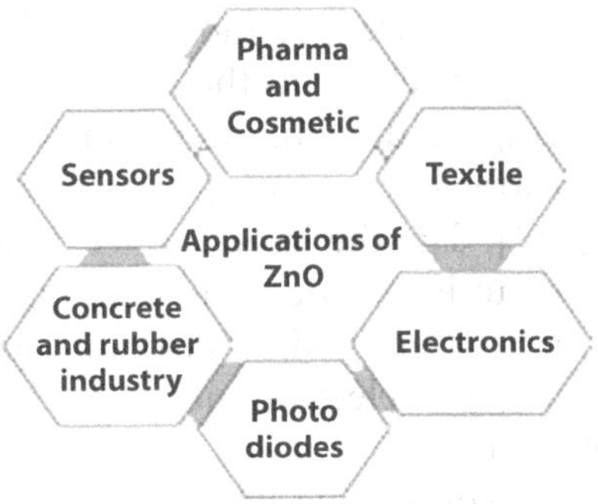

Figure 7.20 Applications of ZnO.

silicone rubber. Zinc oxide nanoparticles can undergo the agglomeration process and as a result, can make larger size particles in the polymer grid of the coatings due to its high surface energy [104]. Thus, for ZnO nanoparticles to inhibit the agglomeration process and to establish the perfect diffusion inside the rubber matrix, surface modification of ZnO is necessary [99, 105].

Textile industry. Zinc oxide has significant surface energy values and vast surface areas, which improve fabric materials' capacity to be attracted to it, increase its desirable role in textile functions, and increase the durability of treated fabrics. Therefore, nanoparticles become attractive materials in textile industry. Various research note that zinc oxide nanoparticles are not only utilized as materials in the finishing processes to improve absorption properties of the treated material in the UV region, but also possess two additional functions: self-cleaning property and antibacterial activity [106].

Photodiodes. Another important application of zinc oxide (ZnO) nanostructures is that it can also be prepared and fabricated in the form of photodiodes. Various researchers reported that ZnO at the nanoscale level can be utilized as photodiodes for scientific and industrial applications [107, 108]. The experimental study of zinc oxide showed that nanostructures of ZnO used as heterojunction photodiodes consist of a p-type silicon and an n-type ZnO nanowire, having a core and shell framework. Various photodiode applications reported that the conformal coating formed by a layer of n-type zinc oxide enclosed with p-type silicon (Si) nanowire is utilized. These photodiodes present improved visible and ultraviolet (UV) responsivity in comparison to other photodiodes with the planar thin film [109, 110].

Sensors. Biosensors such as calorimetric, photometric, piezoelectric, and electrochemical ones, when classified based on the detection principles, are broadly utilized in environmental monitoring, analysis of chemical and biological parameters, healthcare, and industries related to food products [111]. Nanomaterials, separately or in the form of a mixture with bioactive chemical substances, are drawing increasing attention because they can provide a useful platform for growing highly efficient biosensors due to their uniqueness [112], e.g., the large surface area of materials having nanodimensions can be utilized to disable different molecules produced by living organisms such as antibodies, enzymes, and proteins. Further, they can also permit the direct electron movement between biomolecules' active sites and the electrode.

In addition to its semiconducting properties, ZnO materials in nanodimension can also possess different fascinating characteristics for biosensing like strong adsorption capability, good catalytic performance, and large value of the isoelectric point (IEP; ~9.5) that are good enough for adsorption of few proteins (e.g., antibodies having low values of IEPs and enzymes) by electrostatic interactions [113]. Besides this, the excellent properties of the materials, which are suitable for living tissues/stability, have less toxicity, have large surface areas, and have good electron transfer efficiency also show that they can be attracting materials in nanodimensions for biosensor applications [114].

Nanolubricants. Zinc oxide nanolubricants have been widely used since 2010. However, since the 19th century, organic friction modifiers were used as lubricant enhancers as anti-wear and friction modifiers, which are now replaced majorly by nanoparticles. ZnO nanoparticles are most useful among other nanolubricants [96, 97]. Later on, various other nanoparticles have come into existence, which were extensively used as nanolubricants. Some of them are titanium dioxide (TiO_2) and aluminum Oxide (Al_2O_3) [97]. Lubricating oil is the most crucial finding of nanolubricant itself which is the most supportive material in modern industry. For example, nano-ZnO lubricants increase the performance of American Petroleum Institute Synthetic Material (API SM) engine oil, which is evaluated by pressurized differential scanning calorimetry (PDSC), SRV friction and wear test, etc. [115, 116].

Tribology is another term associated with two Greek terms, 'tribos' and 'logos', which mean 'rubbing'. So, the entire science is associated with the controlling and managing wear, friction, and lubrication [116]. The major reasons for the advancement of this tribology are for the prevention of failure of machineries such as failure of engines and bearings. Various industries such as automobiles, aeronautics, and mining, also benefit from this [96, 97, 115, 117]. The second most important reason associated with the same is the prevention of energy loss, which is meant to be a crucial point of discussion as a tribological issue, which originate from high frictions. Around 18% of the total friction loss has been prevented in past 5 to 10 years, which resulted in saving around 174,000 million Euros all over the world [118].

References

1. Wang, Z.L., Zinc oxide nanostructures: Growth, properties and applications. *J. Phys.: Condens. Matter*, 16, 858, 2004.
2. Zhang, Y., Ram, M.K., Stefanakos, E.K., Goswami, D.Y., Synthesis, characterization, and applications of ZnO nanowires. *J. Nanomater.*, 2012, 22, 2012.

3. Ong, C.B., Ng, L.Y., Mohammad, A.W., A review of ZnO nanoparticles as solar photocatalysts: Synthesis, mechanisms and applications. *Renewable Sustainable Energy Rev.*, 81, 536, 2018.
4. Agarwal, H., Kumar, S.V., Rajeshkumar, S., A review on green synthesis of zinc oxide nanoparticles - An eco-friendly approach. *Resour.-Effic. Technol.*, 3, 406, 2017.
5. Manjunatha, R.L., Usharani, K.V., Naik, D., Synthesis and characterization of ZnO nanoparticles: A review. *J. Pharmacogn. Phytochem.*, 8, 1095, 2019.
6. Laurenti, M. and Cauda, V., Porous zinc oxide thin films: Synthesis approaches and applications. *Coatings*, 8, 67, 2018.
7. Znaidi, L., Sol–gel-deposited ZnO thin films: A review. *Mater. Sci. Eng. B*, 174, 18, 2010.
8. Udom, I., Ram, M.K., Stefanakos, E.K., Hepp, A.F., Goswami, D.Y., One dimensional-ZnO nanostructures: Synthesis, properties and environmental applications. *Mater. Sci. Semicond. Process.*, 16, 2070, 2013.
9. Wang, J., Chen, R., Xiang, L., Komarneni, S., Synthesis, properties and applications of ZnO nanomaterials with oxygen vacancies: A review. *Ceram. Int.*, 44, 7357, 2018.
10. Witkowski, B.S., Applications of ZnO nanorods and nanowires — A review. *Acta Phys. Polo A*, 134, 1226, 2018.
11. Segets, D., Gradl, J., Taylor, R.K., Vassilev, V., Peukert, W., Analysis of optical absorbance spectra for the determination of ZnO nanoparticle size distribution, solubility, and surface energy. *ACS Nano*, 3, 1703, 2009.
12. Battez, A.H., Gonzalez, R., Viesca, J.L., Fernandez, J.E., Fernandez, J.M.D., Machado, A., Chou, R., Riba, J., CuO, ZrO and ZnO nanoparticles as antiwear additive in oil lubricants. *Wear*, 265, 428, 2008.
13. Wu, L., Zhang, Y., Yang, G., Zhang, S., Yu, L., Zhang, P., Tribological properties of oleic acid-modified zinc oxide nanoparticles as the lubricant additive in poly-alpha olefin and diisooctyl sebacate base oils. *RSC Adv.*, 6, 69836, 2016.
14. Chikere, C., Faisal, N.H., Lin, P.K.T., Fernandez, C., Zinc oxide nanoparticles modified-carbon paste electrode used for the electrochemical determination of Gallic acid. *Applied Nanotechnology and Nanoscience International Conference, Conf. Series*, Paris, France, p. 012008, 2019.
15. Kumar, S.S., Venkateswarlu, P., Rao, V.R., Rao, G.N., Synthesis, characterization and optical properties of zinc oxide nanoparticles. *Int. Nano Lett.*, 3, 1, 2013.
16. Parihar, V., Raja, M., Paulose, R., A brief review of structural, electrical and electrochemical properties of zinc oxide nanoparticles. *Rev. Adv. Mater. Sci.*, 53, 119, 2018.
17. Sabir, S., Arshad, M., Chaudhari, S.K., Zinc oxide nanoparticles for revolutionizing agriculture: Synthesis and applications. *Sci. World J.*, 2014, 1, 2014.
18. Bacaksiz, E., Parlak, M., Tomakin, M., Ozcelik, A., Karakiz, M., Altunbas, M., The effects of zinc nitrate, zinc acetate and zinc chloride precursors on

investigation of structural and optical properties of ZnO thin films. *J. Alloys Compd.*, 466, 447, 2008.
19. Wang, J., Cao, J., Fang, B., Lu, P., Deng, S., Wang, H., Synthesis and characterization of multipod, flower-like, and shuttle-like ZnO frameworks in ionic liquids. *Mater. Lett.*, 59, 1405, 2005.
20. Wang, Z.L., Splendid one-dimensional nanostructures of zinc oxide: A new nanomaterial family for nanotechnology. *ACS Nano*, 2, 1987, 2008.
21. Chaari, M. and Matoussi, A., Electrical conduction and dielectric studies of ZnO pellets. *Phys. B Condens. Matter*, 407, 3441, 2012.
22. Ozgur, U., Alivov, Y.I., Liu, C., Teke, A., Reshchikov, M.A., Dodan, S., Avrutin, V., Cho, S.-J., Morkoc, H., A comprehensive review of ZnO materials and devices. *J. Appl. Phys.*, 98, 041301, 2005.
23. Bhattacharyya, S. and Gedanken, A., A template-free, sonochemical route to porous ZnO nano-disks. *Microporous Mesoporous Mater.*, 110, 553, 2007.
24. Ludi, B. and Niederberger, M., Zinc oxide nanoparticles: Chemical mechanisms and classical and non-classical crystallization. *Dalton Trans.*, 42, 12568, 2013.
25. Sarkar, S. and Sarkar, R., Synthesis, characterization and tribological study of zinc oxide nanoparticles. *Mater. Today: Proc.*, 44, 3606, 2021.
26. Heris, S.Z., Razbani, M.A., Estelle, P., Mahian, O., Rheological behavior of zinc-oxide nanolubricants. *J. Dispers. Sci. Technol.*, 36, 1073, 2015.
27. Mu, L., Zhu, J., Fan, J., Zhou, Z., Shi, Y., Feng, X., Wang, H., Lu, X., Self-lubricating polytetrafluoroethylene/polyimide blends reinforced with zinc oxide nanoparticles, Hindawi Publishing Corporation. *J. Nanomater.*, 2015, 1, 2015.
28. Scarisoreanu, N., Metai, D.G., Dinescu, G., Epurescu, G., Ghica, C., Nistor, L.C., Dinescu, M., Properties of ZnO thin films prepared by radio-frequency plasma beam assisted laser ablation. *Appl. Surf. Sci.*, 247, 518, 2005.
29. Ni, Y.-H., Wei, X.-W., Hong, J.-M., Ye, Y., Hydrothermal preparation and optical properties of ZnO nanorods. *Mater. Sci. Eng. B*, 121, 42, 2005.
30. Chang, S.-S., Yoon, S.O., Park, H.J., Sakai, A., Luminescence properties of Zn nanowires prepared by electrochemical etching. *Mater. Lett.*, 53, 432, 2002.
31. Ristiac, M., Musiac, S., Ivanda, M., Popoviac, S., Sol–gel synthesis and characterization of nanocrystalline ZnO powders. *J. Alloys Compd.*, 397, L1, 2005.
32. Singh, P., Kumar, A., Deepak, Kaur, D., ZnO nanocrystalline powder synthesized by ultrasonic mist-chemical vapour deposition. *Opt. Mater.*, 30, 1316, 2008.
33. Wang, R.C. and Tsai, C.C., Efficient synthesis of ZnO nanoparticles, nanowalls, and nanowires by thermal decomposition of zinc acetate at a low temperature. *Appl. Phys. A*, 94, 245, 2009.
34. Dinesha, M.L., Prasanna, G.D., Naveen, C.S., Jayanna, H.S., Structural and dielectric properties of Fe doped ZnO nanoparticles. *Indian J. Phys.*, 87, 147–153, 2012.

35. Manjunath, K., Ravishankar, T.N., Kumar, D., Priyanka, K.P., Varghese, T., Naika, H.R., Nagabhushana, H., Sharma, S.C., Dupont, J., Ramakrishnappa, T., Nagaraju, G., Facile combustion synthesis of ZnO nanoparticles using Cajanus cajan (L.) and its multidisciplinary applications. *Mater. Res. Bull.*, 57, 325, 2014.
36. Lee, S.D., Nam, S.-H., Kim, M.-H., Boo, J.-H., Synthesis and photocatalytic property of ZnO nanoparticles prepared by spray-pyrolysis method. *Phys. Proc.*, 32, 320, 2012.
37. Kooti, M. and Sedish, A.N., Microwave-assisted combustion synthesis of ZnO nanoparticles. *J. Chem.*, 2013, 4, 2012.
38. Chandore, V., Carpenter, G., Sen, R., Gupta, N., Synthesis of nano crystalline ZnO by microwave assisted combustion method: An eco friendly and solvent free route. *Int. J. Environ. Sci.:Dev. Monit. (IJESDM)*, 4, 47, 2013.
39. Shetty, A. and Nanda, K.K., Synthesis of zinc oxide porous structures by anodization with water as an electrolyte. *Appl. Phys. A*, 109, 157, 2012.
40. Vijayaprasath, G., Murugan, R., Hayakawa, Y., Ravi, G., Optical and magnetic studies on Gd doped ZnO nanoparticles synthesized by co-precipitation method. *J. Lumin.*, 178, 375, 2016.
41. Vazquez, A., Lopez, I.A., Gomez, I., Growth mechanism of one-dimensional zinc sulfide nanostructures through electrophoretic deposition. *J. Mater. Sci.*, 48, 2701, 2013.
42. Ismail, R.A., Ali, A.K., Ismail, M.M., Hassoon, K.I., Preparation and characterization of colloidal ZnO nanoparticles using nanosecond laser ablation in water. *Appl. Nanosci.*, 1, 45, 2011.
43. Ghaleb, A.-D., Wasan, M.K., Sahib, N.A.V., Preparation and characterization of ZnO nanoparticles by laser ablation in NaOH aqueous solution. *Iran. J. Chem. Chem. Eng.*, 37, 16, 2018.
44. Zeng, H., Cai, W., Li, Y., Hu, J., Liu, P., Composition/structural evolution and optical properties of ZnO/Zn nanoparticles by laser ablation in liquid media. *J. Phys. Chem. B*, 109, 18260, 2005.
45. Yang, L., May, P.W., Yin, L., Scott, T.B., Growth of self-assembled ZnO nanoleaf from aqueous solution by pulsed laser ablation. *Nanotechnology*, 18, 1, 2007.
46. Solati, E., Dejam, L., Dorranian, D., Effect of laser pulse energy and wavelength on the structure, morphology and optical properties of ZnO nanoparticles. *Opt. Laser Technol.*, 58, 32, 2014.
47. Zamiri, R., Zakaria, A., Ahangar, H.A., Darroudi, M., Zak, A.K., Drummen, G.P.C., Aqueous starch as a stabilizer in zinc oxide nanoparticle synthesis via laser ablation. *J. Alloys Compd.*, 516, 41, 2012.
48. Singh, S.C. and Gopal, R., Zinc nanoparticles in solution by laser ablation technique. *Bull. Mater. Sci.*, 30, 291, 2007.
49. O'Brien, P., Chemical vapor deposition, in: *Encyclopedia of Materials: Science and Technology*, Second Edition, p. 1176, 2001.

50. Forleo, A., Francioso, L., Capone, S., Siciliano, P., Lommens, P., Hens, Z., Synthesis and gas sensing properties of ZnO quantum dots. *Sens. Actuators B Chem.*, 146, 111, 2010.
51. Aziz, S.N.Q.A.A., Pung, S.-Y., Ramli, N.N., Lockman, Z., Growth of ZnO nanorods on stainless steel wire using chemical vapour deposition and their photocatalytic activity. *Sci. World J.*, 2014, 1, 2014.
52. Ferreira, M.G.S., Zheludkevich, M.L., Tedim, J., Yasakau, K.A., Self-healing nanocoatings for corrosion control, in: *Corrosion Protection and Control Using Nanomaterials*, p. 213, Woodhead Publishing Series in Metals and Surface Engineering, New Delhi, 2012.
53. Wu, X., Lu, G., Li, C., Shi, G., Room-temperature fabrication of highly oriented ZnO nanoneedle arrays by anodization of zinc foil. *Nanotechnol.*, 17, 4936, 2006.
54. Fox, K.E., Tran, N.L., Nguyen, T.A., Nguyen, T.T., Tran, P.A., Surface modification of medical devices at nanoscale—Recent development and translational perspectives, in: *Biomaterial in Translational Medicine: A Biomaterials Approach*, p. 189, Woodhead Publishing Series in Biomaterials, New Delhi, 2019.
55. Li, A.D. and Liu, W.C., Optical properties of ferroelectric nanocrystal/polymer composites, in: *Physical Properties and Applications of Polymer Nanocomposites*, p. 158, Woodhead Publishing Series in Composites Science and Engineering, New Delhi, 2010.
56. Aneesh, P.M., Vanaja, K.A., Jayaraj, M.K., Synthesis of ZnO nanoparticles by hydrothermal method. *Proc. of SPIE on Nanophotonic Materials IV*, Stefano Cabrini, vol. 6639, 66390, 2007.
57. Brintha, S.R. and Ajitha, M., Synthesis and characterization of ZnO nanoparticles via aqueous solution, sol-gel and hydrothermal methods. *J. Appl. Chem.*, 8, 66, 2015.
58. Rodriguez, C.A.D. and Tremiliosi-Filho, G., Electrochemical deposition, in: *Encyclopedia of Tribology*, Q.J. Wang and Y.W. Chung (Eds.), p. 103, Springer, Boston, MA, Electrochemical Deposition, 2013.
59. Jiangfeng, G., Zhaoming, D., Qingping, D., Yuan, X., Weihua, Z., Controlled synthesis of ZnO nanostructures by electrodeposition method. *J. Nanomater.*, 2010, 1, 2010.
60. Sun, S., Jiao, S., Zhang, K., Wang, D., Li, H., Gao, S., Wang, J., Yu, Q., Guo, F., Zhao, L., Su, S., Morphology and properties of ZnO nanostructures by electrochemical deposition: Effect of the substrate treatment. *J. Mater. Sci.: Mater. Electron.*, 24, 85, 2013.
61. Asmatulu, R., Nanocoatings for corrosion protection of aerospace alloys, in: *Corrosion Protection and Control Using Nanomaterials*, p. 357, Woodhead Publishing Series in Metals and Surface Engineering, New Delhi, 2012.
62. Yilmaz, E. and Soylak, M., Functionalized nanomaterials for sample preparation methods, in: *Handbook of Nanomaterials in Analytical Chemistry*, Modern Trends in Analysis, p. 413, 2020.

63. Vafaee, M. and Ghamsari, M.S., Preparation and characterization of ZnO nanoparticles by a novel sol-gel route, materials. *Mater. Lett.*, 61, 3265, 2007.
64. Gupta, M., Sharma, V., Shrivastava, J., Solanki, A., Singh, A.P., Satsangi, V.R., Dass, S., Shrivastav, R., Preparation and characterization of nanostructured ZnO thin films for photoelectrochemical splitting of water. *Bull. Mater. Sci.*, 32, 30, 2009.
65. Omri, K., El Ghoul, J., Lemine, O.M., Bououdina, M., Zhang, B., El Mir, L., Magnetic and optical properties of manganese doped ZnO nanoparticles synthesized by sol–gel technique. *Superlattices Microstruct.*, 60, 139, 2013.
66. El Ghoul, J., Barthou, C., Saadoun, M., El Mir, L., Synthesis and optical characterization of SiO/Zn SiO:Mn nanocomposite. *J. Phys. B*, 405, 597, 2010.
67. Alwan, R.M., Kadhim, Q.A., Sahan, K.M., Ali, R.A., Mahdi, R.J., Kassim, N.A., Jassim, A.N., Synthesis of zinc oxide nanoparticles via sol–gel route and their characterization. *Nanosci. Nanotechnol.*, 5, 6, 2015.
68. Itskos, G., Nikolopoulos, N., Kourkoumpas, D.-S., Koutsianos, A., Violidakis, I., Drosatos, P., Grammelis, P., Energy and the environment, in: *Environ. Dev.*, Basic Principles, Human Activities, and Environmental Implications, p. 363, 2016.
69. Raoufi, D., Synthesis and microstructural properties of ZnO nanoparticles prepared by precipitation method. *Renewable Energy*, 50, 932, 2013.
70. Hutera, B., Kmita, A., Olejnik, E., Tokarski, T., Synthesis of ZnO nanoparticles by thermal decomposition of basic zinc carbonate. *Arch. Metall. Mater.*, 58, 489, 2013.
71. Ravichandran, K., Praseetha, P.K., Arun, T., Gobalakrishnan, S., Synthesis of nanocomposites, in: *Synthesis of Inorganic Nanomaterials*, Advances and Key Technologies, Micro and Nano Technologies, p. 141, 2018.
72. Khan, I., Khan, S., Nongjai, R., Ahmed, H., Khan, W., Structural and optical properties of gel-combustion synthesized Zr doped ZnO nanoparticles. *Opt. Mater.*, 35, 1189, 2013.
73. Liu, H.R., Shao, G.X., Jia, W., Zhang, Z.X., Zhang, Y., Liang, J., Liu, X.G., Jia, H.S., Xu, B.S., Morphological evolution of ZnO microspheres from Zn (OH) Ac·2H O by ultrasonic irradiation method. *CrystEngComm*, 15, 3615, 2013.
74. Patnaik, P., *Dean's analytical chemistry handbook*, 2nd eds., p. 1114, McGraw-Hill, New York, U. S., 2004.
75. Vijayaprasath, G., Murugan, R., Asaithambi, S., Sakthivel, P., Mahalingam, T., Hayakawa, Y., Ravi, G., Structural and magnetic behavior of Ni/Mn co-doped ZnO nanoparticles prepared by co-precipitation method. *Ceram. Int.*, 42, 2836, 2016.
76. Huang, J., Lin, L., Sun, D., Chen, H., Yang, D., Li, Q., Bio-inspired synthesis of metal nanomaterials and applications. *Chem. Soc. Rev.*, 44, 6374, 2015.
77. Hakim, L.F., Portman, J.L., Casper, M.D., Weimer, A.W., Aggregation behavior of nanoparticles in fluidized beds. *Powder Technol.*, 160, 149, 2005.
78. Mukherjee, P., Ahmad, A., Mandal, D., Senapati, S., Sainkar, S.R., Khan, M.I., Parishcha, R., Ajaykumar, P.V., Alam, M., Kumar, R., Sastry, M.,

Fungus-mediated synthesis of silver nanoparticles and their immobilization in the mycelial matrix: A novel biological approach to nanoparticle synthesis. *Nano Lett.*, 1, 515, 2001.
79. Tripp, S.L., Pusztay, S.V., Ribbe, A.E., Wei, A., Self-assembly of cobalt nanoparticle rings. *J. Am. Chem. Soc.*, 124, 7914, 2002.
80. Vidya, C., Hiremath, S., Chandraprabha, M.N., Antonyraj, M.A.L., Gopal, I.V., Jain, A., Bansal, K., Green synthesis of ZnO nanoparticles by Calotropis Gigantea. *Int. J. Curr. Eng.*, 1, 118–120, 2013.
81. Dadalioglu, I. and Evrendilek, G.A.J., Chemical compositions and antibacterial effects of essential oils of Turkish oregano (Origanum minuti-florum), bay laurel (Laurus nobilis), Spanish lavender (Lavandula stoechas L.), and fennel (Foeniculum vulgare) on common foodborne pathogens. *Agric. Food Chem.*, 52, 8255, 2004.
82. Fakhari, S., Jamzad, M., Fard, H.K., Green synthesis of zinc oxide nanoparticles: A comparison. *Green Chem. Lett. Rev.*, 12, 19, 2019.
83. Suresh, J., Pradheesh, G., Alexramani, V., Sundrarajan, M., Hong, S.I., Green synthesis and characterization of zinc oxide nanoparticle using insulin plant (Costus pictus D. Don) and investigation of its antimicrobial as well as anticancer activities. *Adv. Nat. Sci.: Nanosci. Nanotechnol.*, 9, 015008, 2018.
84. Santhoshkumar, J., Venkat, K.S., Rajeshkumar, S., Synthesis of zinc oxide nanoparticles using plant leaf extract against urinary tract infection pathogen. *Resour. Effic. Technol.*, 3, 459, 2017.
85. Singh, A.K., Experimental methodologies for the characterization of nanoparticles. Structure, properties and mechanisms of toxicity, in: *Engineered Nanoparticles*, p. 125, 2016.
86. Raut, S., Thorat, P.V., Thakre, R., Green synthesis of zinc oxide (ZnO) nanoparticles using ocimum tenuiflorum leaves. *Int. J. Sci. Res.*, 4, 1225, 2015.
87. Marouf, S., Beniaiche, A., Guessasa, H., Azizib, A., Morphological, structural and optical properties of zno thin films deposited by dip coating method. *Mater. Res.*, 20, 95, 2017.
88. Tauc, J. and Menth, A., Corelation between the solution chemistry to observed properties of CdTe thin films prepared by CBD method. *J. Non Cryst. Solids*, 8-10, 569, 1972.
89. Shan, F.K. and Yu, Y.S., Band gap energy of pure and Al-doped ZnO thin films. *J. Eur. Ceram. Soc.*, 24, 1869, 2004.
90. Talam, S., Karumuri, S.R., Gunnam, N., Synthesis, characterization, and spectroscopic properties of ZnO nanoparticles. *ISRN Nanotechnol.*, 2012, 1, 2012.
91. Wang, X., Ding, Y., Summers, C.J., Wang, Z.L., Large-scale synthesis of six-nanometer-wide ZnO nanobelts. *J. Phys. Chem. B*, 108, 8773, 2004.
92. Huang, M.H., Wu, Y., Feick, H., Tran, N., Weber, E., Yang, P., Catalytic growth of zinc oxide nanowires by vapor transport. *Adv. Mater.*, 13, 116, 2001.

93. Williams, G. and Kamat, P.V., Graphene–semiconductor nanocomposites:- Excited-state interactions between ZnO nanoparticles and graphene oxide. *Langmuir*, 25, 13869, 2009.
94. Zhou, J., Zhao, F., Wang, Y., Zhang, Y., Yang, L., Size-controlled synthesis of ZnO nanoparticles and their photoluminescence properties. *J. Lumin.*, 122-123, 195, 2007.
95. Rao, B.S., Kumar, B.R., Reddy, V.R., Rao, T.S., Preparation and characterization of CdS nanoparticles by chemical co-precipitation technique. *Chalcogenide Lett.*, 8, 177, 2011.
96. Elagouz, A., Ahmed Ali, M.K., Xianjun, H., Abdelkareem, M.A.A., Hassan, M.A., Frictional performance evaluation of sliding surfaces lubricated by zinc-oxide nano-additives. *Surf. Eng.*, 36, 144, 2019.
97. Singh, A., Chauhan, P., Mamatha, P.G., A review on tribological performance of lubricants with nanoparticles additives. *Mater. Today: Proc.*, 25, 586, 2020.
98. Kołodziejczak-Radzimska, A. and Jesionowski, T., Zinc oxide—From synthesis to application: A review. *Materials*, 7, 2833, 2014.
99. Matei, A., Tucureanu, V., Dumitrescu, L., Aspects regarding synthesis and applications of ZnO nanomaterials. *Bull. Transilv. Univ. Bras.*, 7, 45, 2014.
100. Kundu, T.K., Karak, N., Barik, P., Saha, S., Optical properties of ZnO nanoparticles prepared by chemical method using poly (Vinyl Alcohol) (PVA) as capping agent. *Int. J. Soft Comput. Eng.*, 1, 19, 2011.
101. Verma, P., Pandey, S., Pandey, A.C., Synthesis and photoluminescence of ZnO nanophosphors. *Proc. ASID*, New Delhi, p. 256, 2006.
102. El-Diasty, E.M., Ahmed, M.A., Okasha, N., Mansour, S.F., El-Dek, S.I., El-Khalek, H.M.A., Youssif, M.H., Antifungal activity of zinc oxide nanoparticles against dermatophytic lesions of cattle. *Rom. J. Biophys.*, 23, 191, 2013.
103. Singh, P. and Nanda, A., Antimicrobial and antifungal potential of zinc oxide nanoparticles in comparison to conventional zinc oxide particles. *J. Chem. Pharm. Res.*, 5, 457, 2013.
104. Momen, G. and Farzaneh, M., Survey of micro/nano filler use to improve silicone rubber for outdoor insulators. *Rev. Adv. Mater. Sci.*, 27, 1, 2014.
106. Kathirvelu, S., D'Souza, L., Dhurai, B., UV protection finishing of textiles using ZnO nanoparticles. *Indian J. Fibre Text. Res.*, 34, 267, 2009.
107. Marie, M., Mandal, S., Manasreh, O., An electrochemical glucose sensor based on zinc oxide nanorods. *Sensors*, 15, 18714, 2015.
108. Kim, K.H., Utashiro, K., Abe, Y., Kawamura, M., Structural properties of zinc oxide nanorods grown on al-doped zinc oxide seed layer and their applications in dye-sensitized solar cells. *Materials*, 7, 25322, 2014.
109. Singh, S.N., Kumari, S., Das, B.K., Electrical properties of polycrystalline silicon and zinc oxide semiconductors. *Bull. Mater. Sci.*, Springer, 6, 243, 1984.
110. Park, E., Lee, J., Park, T., Lee, J., Lee, D., Sung, M.M., Yi, W., Aqueous synthesis of n-/p-type ZnO nanorods on porous silicon for the application of p-n junction device. *10th IEEE Conf. Nanotech.*, Seoul, South Korea, p. 977, 2010.

111. Zhao, Z., Lei, W., Zhang, X., Wang, B., Jiang, H., ZnO-based amperometric enzyme biosensors. *Sensors*, 10, 1216, 2010.
112. Yakimova, R., Selegard, L., Khranovskyy, V., Pearce, R., Sptez, A.L., Uvdal, K., ZnO materials and surface tailoring for biosensing. *Front. Biosci.*, 4, 278, 2012.
113. Wang, J.X., Sun, X.W., Wei, A., Lei, Y., Cai, X.P., Li, C.M., Zinc oxide nanocomb biosensor for glucose detection. *Appl. Phys. Lett.*, 88, 233106, 2006.
114. Kumar, S.A. and Chen, S.M., Nanostructured zinc oxide particles in chemically modified electrodes for biosensor applications. *Anal. Lett.*, 41, 58, 2008.
115. Weiguo, X., Zhenghua, Z., Peng, W., Zhiliang, J., Xiaohong, X., Xugaung, Z., Performance study of zinc oxide nanoparticles for lubricant oil. *Adv. Mat. Res.*, 1118, 195, 2015.
116. Shahnazar, S., Bagheri, S., Hamid, S.B.A., Enhancing lubricant properties by nanoparticle additives. *Int. J. Hydrog. Energy*, 41, 3153, 2016.
117. Esfe, M.H., Goodarzi, M., Reiszadeh, M., Afrand, M., Evaluation of MWCNTs-ZnO/5W50 nanolubricant by design of an artificial neural network for predicting viscosity and its optimization. *J. Mol. Liq.*, 277, 921, 2019.
118. Kedzierski, M.A., Brignoli, R., Quine, K.T., Brown, J.S., Viscosity, density, and thermal conductivity of aluminum oxide and zinc oxide nanolubricants. *Int. J. Refrig.*, 74, 3, 2017.

8

Improvement in the Properties of Biodegradable Nanolubricants

Sandip Paul Choudhury[1]*, Pushpendra Singh Shekhawat[1], Debanjan Bhattacharjee[2] and Umesh K. Dwivedi[1]

[1]*Amity School of Applied Science, Amity University, Jaipur, Rajasthan, India*
[2]*Department of Physics, National Institue of Technology (NIT) Agartala, Tripura, India*

Abstract
Over the years, the application of lubricants has been emerging and is potentially essential in most machineries, but due to the hazards conventional lubricants have, a paradigm shift towards biodegradable nanolubricants has been called for. In the process, researchers have worked towards improving the tribological properties of biodegradable nanolubricants use, varied base material, and additives. The present chapter discusses the types of biodegradable nanolubricants and how to improve their properties.

Keywords: Lubricants, biodegradable nanolubricants, additives

8.1 Introduction

The technology of lubricants is considered one of the most useful and oldest technologies. Lubricants are used between two surfaces. They can be liquid, solid, or gas, which acts as an extra layer and helps in avoiding the roughness of the layers around it. A lot of mechanical systems as per their functionality require different lubricants so that the required energy for the operation can be minimized by lowering the contact friction. The mechanisms behind the reduction of friction include the mending effect, third

**Corresponding author*: sandip.pchoudhury@gmail.com

Mohd Yusuf, Lalit Prasad and Shafat Ahmad Khan (eds.) Nanolubricants: Generation and Applications, (209–228) © 2024 Scrivener Publishing LLC

body material transfer, the formation of tribofilms, the rolling effect, and the protective film formation [1]. There are three broad types of lubricating oils: vegetable, mineral, and synthetic. Mineral oils are commonly utilized as lubricating oils in industries.

8.1.1 Why Biodegradable Lubricants?

Due to their poor degradability, most lubricants used in the industry cause environmental risks when they are spread. Environmental health issues and the rise in pollution have recently become concerns for every nation. The leakage of lubricants into the environment is a major worry in the equipment lubrication industry. Failures in the lubricating system immediately affect the soil and water, whereas volatile lubricants contaminate the air. Growing environmental consciousness has acted as a catalyst for lubricating technology advancement. So, in regions where the environment needs to be protected, the application of biodegradable lubricants is inevitable. As an efficient replacement for petroleum-based oils, biodegradable oils are employed since they are non-toxic, renewable, biodegradable, and environmentally beneficial. Additionally, they possess crucial lubricant qualities such as low volatility, elevated viscosity index, a superb solvent for additives, and outstanding lubricity [2].

8.1.2 Vegetable Oil-Based Lubricants

Before the discovery of petroleum, vegetable oils were frequently utilized as lubricants in transportation vehicles and machinery. Petroleum has dominated the lubricant industry because it is more cost-effective and performs better. However, vegetable oil as a lubricant has been recently explored more because of legal issues and shifting crude oil prices. This could boost markets for unusual seeds and their oil, hence raising farmer incomes and maximizing the use of agricultural goods. Vegetable oil, when used as lubricants, has merits and demerits. In many industries, vegetable oil-based lubricants are used as an alternative to synthetic oils because of their non-toxic and environment-friendly nature [3]. Due to the triacylglycerol structure having longer fatty acid chains and the existence of polar groups, such oils possess an amphiphilic nature, and they have become a very good choice as lubricants. Surface film is enhanced by the arrangement of molecules in closed-packed monomolecular surfaces due to the triacylglycerol structures present in vegetable oils, and better viscosity properties and low volatility are provided by high molecular weight. However, vegetable oils in their natural form are not suitable for industrial uses due to their bisallylic nature and poor thermos-oxidation. When the oils oxidize, acid content and viscosity increase, resulting in reduced performance of machines due to sludge formation and corroding parts. Researchers have also concluded that many vegetable oils experience solidification, haziness, precipitation, and degraded flow rates under harsh environmental, pressure, and temperature conditions. This limitation creeps in when vegetable oils are used as lubricants. Vegetable-based lubricants' intrinsic poorer physicochemical qualities compared to those derived from mineral oil, are what restrict their performance [4].

8.1.3 Synthetic Lubricants

The industry has been told by strict new environmental standards to cut back on its use of finite natural resources, industrial waste, and dangerous pollutants. While a significant number of applications need lubrication on the one hand, consumption needs to be decreased on the other. For applications where mineral base oil could not work as desired, such as in extreme wear conditions, and where requirements of properties like better shelf life, better equipment efficiency, or being less prone to burning were required, synthetic lubricants were explored and utilized. Synthetic lubricants are helpful as they are cost-effective in areas like machine longevity, oil usage, energy use, environmental support, and security. Therefore, their usage is not just limited to specialized applications. Situations where synthetic oils have clear benefits compared to conventional lubricants include a major extension of oil-change interval, fewer machine failures, and environmental safety, among many other successful applications. Synthetic base stocks are created artificially, have a regulated molecular structure, and have known qualities as opposed to mineral-based oil stocks, which are just a complicated blend of naturally formed hydrocarbons extracted from raw oil. Synthetic oil is created by a combination of base stocks and selected additives, just like any modern fluid lubricant. A specific synthetic lubricant is frequently chosen for a particular application based on one or more outstanding qualities that set it apart from traditional petroleum oils and/or applications where the extra cost can be justified. When combined with additives, synthetic oil technology enables the materials to be tailored for specific applications where lubricant is required and offers targeted performance. Mineral oil and synthetic lubricants are commonly used together to get the necessary property combinations. Synthetic lubricants have several benefits over oils extracted from minerals, including low-temperature fluid flow (which improves performance in cold weather), low volatility elevated temperature stability (elevated "viscosity index"), low oxidation, and high inbuilt detergent behavior (which helps to keep the engine clean with minimum additives in it). The environmental benefits of synthetic oils include better fuel efficiency, less oil usage, and longer intervals between oil changes. The reactive and material qualities of the basic fluids and the impact of the additives are used to determine the features of synthetic lubricants. Synthetic lubricants frequently contain additives, which have a stronger impact on their performance, affecting their oxidation behavior, load-holding capacity, rust prevention, foaming, and demulsification. Environmentally adapted lubricants (EALs), one of the newly developed goods, differ from traditional mineral oil-based products in terms of their fundamental features.

8.1.4 Properties and Synthesis of Nanolubricants

Lubricants, in general, are selected based on application and its properties. Some of the important properties are flash point, viscosity, fire point, pour point, and oxidation behavior. Viscosity is an important criterion for choosing the right lubricant. The oil structure, which includes the carbon chain length, decides the material viscosity. More hydrocarbon chains will result to more viscosity. Viscosity is also a temperature-dependent factor measured by the viscosity index. A lubricant will be more stable if it has a higher viscosity index. The lowest temperature at which vapor is produced by heating the lubricant is the flash point. The minimum temperature at which continuous ignition happens is the fire point. An ideal lubricant should have more fire point than the operational temperature of the device. The minimum temperature at which the oil flows is known as the pour point. Although vegetable oil has a wide application, a lower oxidation temperature is its major drawback. These parameters can determine the appropriate lubricant for specific applications. But at the same time, the improvement in these properties is also a requirement to optimize a system.

Nanolubricants can be synthesized using one-step methods like vapor deposition, microwave, chemical methods, laser ablation, polyol process, physical vapor condensation, plasma technique, and arc synthesis. Two-step methods for the synthesis of nanolubricants can include microwave synthesis and mixing. For the synthesis and preparation of nanolubricants, additives like defoamers, antioxidants, corrosion inhibitors, emulsifiers, and extreme agent agents are utilized based on the application. The base material may be water or a lubricant whose properties are modified by the addition of nanoparticles with the help of a dispersant/surfactant. Appropriate dispersal methods like ultrasonic treatment and mechanical agitation are used (Figure 8.1).

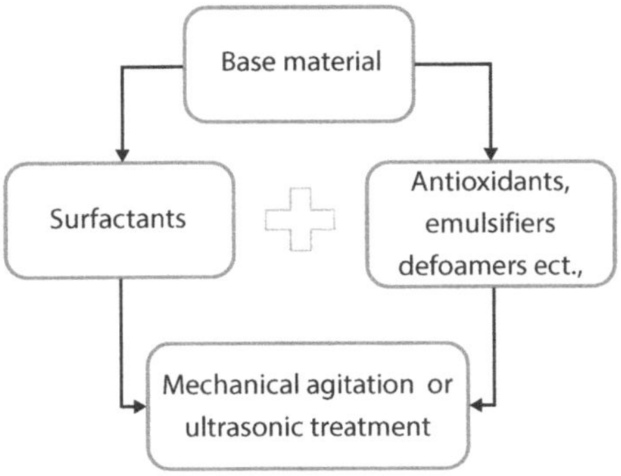

Figure 8.1 Synthesis of biodegradable nanolubricants.

The vapor deposition method is one of the most common methods for synthesizing nanolubricants. As shown in Figure 8.1, the base material and the nanoparticles are directly mixed in the presence of a surfactant. After the heating and cooling process, the nanolubricants are finally collected as a fluid. In such a one-step method, there may be impurities, which may still be left in the final product due to an incomplete reaction. The two-step methods listed above can be utilized to avoid this disadvantage. In such methods, additional treatments such as mechanical stirring, high-shear mixing, and magnetic force agitation are used.

8.2 Nanoparticles for Lubricants

In the past 10 years, the utilization of nanoparticles in various types of base fluids has drawn attention. For instance, water having dispersing nanoparticles in it could improve its thermal conductivity and make it a superior heat-transfer fluid, particularly for applications involving solar collectors. Since the early 2000s, genuine uses for nano-coolants—dispersions of nanoparticles in conventional coolants such as ethylene glycol and water—have been studied. It is anticipated that adding nanoparticles to a lubricant will give it anti-wear and anti-friction capabilities by creating a protective film and employing a rolling effect between the layers of the lubricants, respectively (Figure 8.2).

The friction behavior and load-holding nature of the friction components in physical systems are both significantly reduced when nanoparticles are added to lubricating oil. The increase in lubrication of the lubricating oil enhanced with nanoparticles has been explained by several

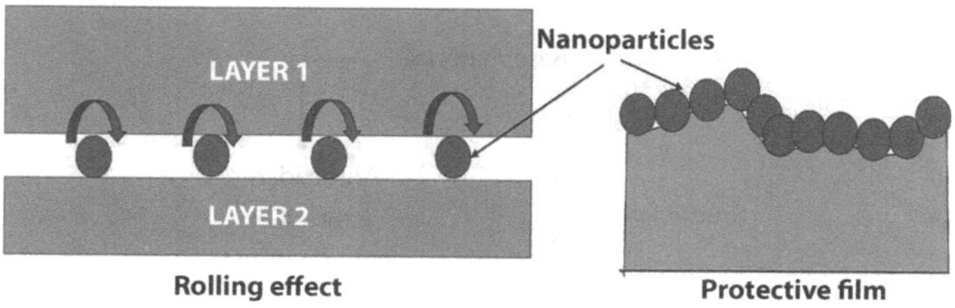

Figure 8.2 Rolling effect and protective film formation by using nanoparticles.

mechanisms, such as the ball-bearing effect, safety coating, joining effect, and smoothing effect. These processes can be divided primarily into two categories. The first category is how the nanoparticles themselves directly enhance lubrication. Between the friction surfaces, the nanoparticles floating in the lubricant act as ball bearings. By coating the rough surfaces, they provide a safety film. The second category involves surface augmentation due to the presence of nanoparticles. The joining effect is the process by which the additive accumulates on the friction surface and makes up for mass loss [5]. Additionally, the polishing action of nanoparticle-assisted abrasion reduces the graininess of the lubricating region. Furthermore, it is proven that inserting the right number of nanoparticles into the lubricating material has well-enhanced its anti-friction and anti-wear properties [6]. However, creating stable suspension-based nanoparticles is the key problem since less stable suspensions of lubricants over extended periods of time in stationary conditions would cause agglomerates and sediments to form. Therefore, compatibility between the main oil and nanoadditives is a crucial criterion that must be taken into consideration to ensure the production of an efficient suspension or nanolubricant. In addition, because of the large size of the particles that tend to accumulate in lubricants during operation, as well as the presence of oxygen and humidity, the relatively poor tribological qualities of nanolubricants become a limiting issue in practical implementation [7]. Numerous studies have been done on the inclusion of nanoparticles ranging in size from 2 to 120 nm, which function as smoothening agents and significantly lessen friction and wear. All the studies have made use of a variety of nanoparticles, including SiO_2, metals and their oxides, metal borate, metal sulfide, and carbonate-based metals. They all concurred that the shape, size, concentration, and degree of crystallinity (defect) all influenced the nanolubricant's tribological performance.

Compared to organic compounds typically utilized as lubricant additives, nanoparticles provide several significant advantages:

- They can easily infiltrate the contact region due to their size in the nano range.
- They frequently function well at room temperature; therefore, there is no requirement for an induction period to achieve important tribological features.

Nanoparticles can be used in lubricants in various ways, which include lubrication by micellar systems, metallic nanostructures, and coated colloids.

Micelles are agglomerations of surface-active agent molecules and are formed by detergents in oils (Figure 8.3). A polar part and a nonpolar, hydrophobic part make up polar compounds. They can create reverse micelles or micelles depending on the polarity of the solvent. By forming a steric barrier, hydrocarbon chains stop the particles from aggregating. Micelles that are overbased have a mineral core, such as calcium carbonate ($CaCO_3$). They can simply dissolve in the main oil, and the dispersion remains even for an extended period due to their lipophilic linkages and nanosize. Initially employed as detergents, overbased micelles also have anti-wear qualities. A thin coating applied to a crystalline surface, a dispersion of colloids in a polar medium (grease, oil etc.), and gas state lubrication are only a few applications for overbased micelles. Most of the time, there exists a polyphasic complex reaction that produces overbased micelles. They are made by carbonating a metal or by using hydroxide that has been dissolved in a polar combination of surface-active agents, aromatic

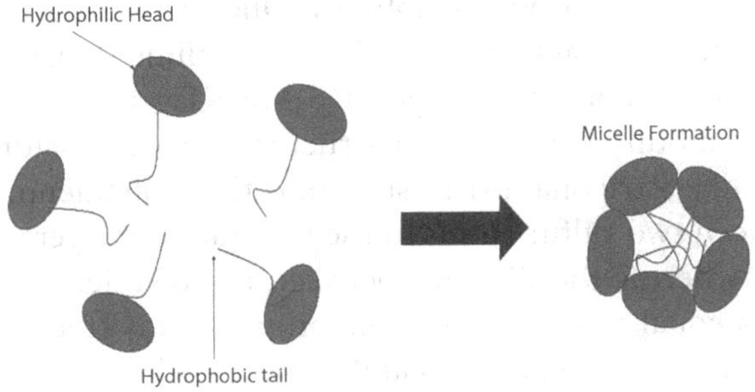

Figure 8.3 Micelle formation due to surfactants.

solvents, and paraffinic oil. Thus, spherical colloidal particles between 2 and 20 nm in size can be created from oxide or hydroxide particles [8, 9].

While soft metals like Ag, Au, Cu, Pb, Zn, Bi, In, and their alloys are employed as lubricants in the form of surface coatings and films, rigid metals or alloys, like steel, are utilized to make tribo-pairs for rolling or sliding bearings. The face-centered geometry, which contains numerous slip-planes and, thus, displays lower shear strength all along the sliding direction, is what causes these soft metals' lubricating behavior. At low friction speeds, lubricating films made of soft metals may exhibit some self-repairing characteristics. However, in most cases, they do have the following disadvantages: (a) they do not exhibit this behavior at greater sliding speeds and/or loads; (b) they are unable to dissipate friction heat; (c) relubrication is typically not an option; and (d) application procedures are frequently difficult. Metal particles between 1 and 100 nm in size are referred to as metal nanoparticles. Whether they are powders or thin films, these materials' sizes are what give them their unique features. Metals, whether they are in the form of powders or thin films, will display amazing and unexpected features when at least a single dimension is in the nanometer range, which is not present in their bulk phase. The synthesis, structure characterization, characteristics, and applications of metal nanoparticles have all been the subject of an extensive investigation. In tribology, metal nanoparticles are anticipated to be effective lubricants when added to lubricating material for the reasons listed: (a) because nanoparticles are minute, an appropriate colloidal dispersion in the lubricant can be accomplished through correct methods, avoiding the undesirable process caused by gravitation and (b) because such nanoparticles are more likely to be captured in the surface undergoing friction with the generation of a stable, well-distributed dispersion [10].

Nanoparticles can be incorporated into coatings or utilized as coatings themselves. Coatings with intriguing properties can be made by combining IF-MS2, C60, and carbon nanotubes. Due to its intriguing friction-reducing characteristics associated with its crystalline structure, MoS_2 has been used as a lubricant for many years. S-Mo-S layers are stacked to create MoS_2, a compound with lamellar structure. Six sulfur atoms at the tip of something like a trigonal prism surround the molybdenum atom. The distance between two sulfur atoms in the two adjacent layers is equivalent to 0.349 nm, whereas the distance between a molybdenum atom and a sulfur atom is equal to 0.241 nm. This property was frequently cited to explain the ease of layer separation and, hence, the lubricating qualities of MoS_2 [10]. The efficiency of the MoS_2 layers depends on the environment in which they are utilized. The coefficient of friction rises, with humidity up to 65% RH (relative humidity) when two metal surfaces coated with MoS_2 come into contact, and then it decreases [11]. The friction parameter reduces with improved proximity pressure or sliding velocity. The water included in MoS_2 films would evaporate at high proximity pressures or fast sliding speeds due to the elevated temperature of the touching area [12]. As a result, the water content is mostly responsible for the film's longevity.

The substitution of conventional lubricants with biodegradable nanolubricants is a necessity considering the environmental impact and health concerns. But biodegradable oils in use in its pure form have some disadvantages including low temperature applications [13]. But as discussed, their properties can be significantly improved by use of additives. Tribological improvement can be achieved by adding carbon-based structures, metals, etc., to the host lubricants [14–16]. The improvement in the properties is studied by an analysis of changes in microstructure, surface repairing effect, tribofilm formation, etc. [16–20].

8.3 Types of Biodegradable Nanolubricants

A lubricant is more useful and acceptable if it is environmentally neutral. A lubricant must be eco-friendly at the manufacturing or fabricating level because it uses less energy with minimum waste materials, which has harmful emission. The lubricant can be considered eco-friendly if it has the listed properties:

- The lubricant should have a biodegradable nature.
- The oil must be non-toxic and non-carcinogenic.
- The quick biodegradability of the oil should not create any harmful decay products.
- The main lubricating oil must be extracted from a renewable source without any depletion of natural resources and with a negligible greenhouse effect.
- The lubricant should be ecotoxicologically acceptable.

Biodegradable nanolubricants can be categorized as highly double-bonded or high oleic vegetable oils, low viscosity poly alpha olefins, polyalkylene glycols, acid esters, and polyol esters. Biodegradable nanolubricants can be broadly divided into vegetable oil-based, water-based, and additives-based biodegradable nanolubricants.

8.3.1 Vegetable Oil as a Biodegradable Lubricant

Using vegetable oil as a lubricant is a viable option as it is biodegradable, renewable, and eco-friendly in nature. Vegetable oils are triacylglycerides, synthesized by the esterification of three hydroxyl groups of the glycerol fragment with carboxyl groups of fatty acids. Vegetable lubricants are biodegradable in nature due to the existence of esters [21]. Natural esters can degrade with or without oxygen [22]. The biodegradation rate is dependent on the structure of the base oil and additives. The degree of biodegradation is based on the following conditions [23, 24]:

- Constituents like amino acids and fatty acids are biodegradable.
- Compounds like aromatic compounds, halogenated compounds, nitro compounds, and sulfonated compounds are resistant to biodegradation.
- Straight carbon chain compound biodegrades more actively than branched compounds.
- The biodegradation process generally depends on the position and also the degree of chain branching.
- The biodegradability of transesterified vegetable oil decreases with acyl and alcohol chain length.

The non-toxic and eco-friendly properties of lubricants like vegetable oil make them good substitutes for other industrial lubricants (Table 8.1). Owing to the existence of polar groups as components of the oil, vegetable oil's structural properties such as long fatty acid chains and triacylglycerol structure add to its biodegradable characteristics. The molecules of such triacylglycerol structures form a close-packed monomolecular layer, which in turn improves the surface film, reduces the volatility, and improves the viscosity. However, the naturally available form of the vegetable oil has poor thermal oxidation due to bisallylic protons. This leads to the unsuitability of pure vegetable oil for industrial applications. The reduction in efficiency of pure vegetable oil can be attributed to oil oxidation, resulting in increased acid content and oil viscosity. Such process corrodes metal parts and generates sludge. Some vegetable oils can also undergo cloudiness, precipitation, irregular flow under extreme conditions such as temperature, pressure and environment, and solidification [25–27]. Mineral oil has superior physicochemical characteristics compared to vegetable lubricants, which is a major drawback restricting the use of vegetable oil. Recent research work focuses on the improvement and overcoming these drawbacks to synthesize better biodegradable lubricants.

Table 8.1 Lubricants (vegetable oil) with industry applications [28].

Vegetable oil	Application
Canola oil	Metal fluids and hydraulic oil tractor transmission fluids
Castor oil	Greases and gear lubricants
Olive oil	Lubricants for the automobile industry
Palm oil	Greases and rolling lubricants
Rapeseed oil	Greases, farm equipment (air compressor), chain lubricants
Safflower oil	Enamels, diesel, color paints
Linseed oil	Stains, coating, lacquers, paints
Soybean oil	Shampoo, lubricants, soaps, biodiesel fuel, pesticides, metal casting, hydraulic oil
Jojoba oil	Greases, lubricant application, cosmetics
Crambe oil	Chemicals, surfactants, greases
Sunflower oil	Diesel fuel, greases

8.3.2 Additives-Based Biodegradable Nanolubricants

Elaborate research is being carried out on the vegetable lubricants to enhance their properties. Modification with appropriate additives is a popular technique to improve the vegetable oil's features like anti-wear, oxidation stability, anti-corrosion, anti-friction, and stability against degradation by biological methods. The use of additives is based on the application, as listed in Table 8.2 [29].

Joseph and Sharma attempted to reduce the oxidation of vegetable oils from India by using antioxidants [30]. The stability of the vegetable oil was improved by using amine and phenol-type antioxidants. The viscosity change of soybean oil was delayed by using antioxidants as studied by Ruger *et al.* [31]. The temperature was adjusted to 105°C and 10 mg of iron and copper, which are colloidal metals, were added for the oil testing. This simulated the real operating conditions of gasoline engine crankcase. Their findings tallied with those of Pokorny [29], whose work showed that the TBHQ (tertiary butylhydroquinone) can help to stabilize the soybean viscosity. The antioxidant concentration, which is optimum, was found to be 0.01%. This is similar to the quantity used in food applications. High pressure (HP) and

Table 8.2 Popular lubricant additives and their impact on lubricants.

Category	Additive	Effects
Deposit control additives	Antioxidants	Antioxidants' improvement, radical scavenging, and peroxide inhibition
	Zinc Dithiophosphates (ZDDP)	Improvement of anti-wear and antioxidant properties
	Ash less phosphorus	Improvement of anti-wear properties compared to ZDDP by the creation of thinner and smoother films
	Detergents	Acid neutralization, which damages the metal surface
	Dispersants	Non-polar molecules will be suspended in a lubricant.
Film forming additives	Solid lubricants as friction minimizers	Replacement of a solvent lubricant as it improves friction between surfaces.
	Organic friction modifiers	Better lubricity, energy conservation and minimized friction.
AW and EP additives	Ash less AW and EP additive sulfur carriers	To improve AW and EP
Viscosity control additives	Viscosity modifier using Olefin copolymer (OCP)	High efficiency and affordable for lubrication thickening.
	Viscosity modifier using polymethacrylate (PMA) and pour point depressants (PPD)	Lubricant deposition and wax formation are controlled by PPD. PMA improves viscosity index (VI).
	PPD	Lubricant deposition and wax generation are controlled.

anti-wear (AW) are two main properties that need to be improved when using additives in vegetable oil. When such properties are modified, the frictional impact and wearing capacity will be comparatively higher on mixed/boundary lubricating situation and under heavy loads. Available vegetable oils such as Jatropha and palm oil showed larger friction parameters and wear scar diameters in comparison to other available stamping oils [32]. The effectiveness of vegetable oil-based lubricants (coconut, canola, and sunflower oils) with extreme pressure (EP) components in machining was evaluated in a recent intriguing study by Kumar *et al.* [31]. According to their findings, coconut oil with a 10% EP additive is the best as a cutting fluid in comparison to other vegetable oils and EP additive concentrations.

Because these additions tend to form layers of readily sheared chlorides, sulfides, or phosphides on the metal surface, which essentially avoid extreme wear and damage, the classic HP and AW additives contain chlorine, sulfur, and phosphorous. However, the majority of additives that work best with mineral-based oils do not work well with vegetable lubricants, whereas compounds including sulfur, chlorine, and phosphorus have been avoided since they are detrimental to human health and the surroundings [33]. In addition, to improve oil qualities, epoxidation, transesterification, post-esterification, and hydrogenation are chemical modification procedures that can be used to enhance the physicochemical properties of vegetable oils [34, 35].

According to the results of various studies, transesterification of vegetable oil can enhance its viscosity, lubricity, oxidative stability, and thermal stability by removing the carbon from the ester chain. Several research has shown that methyl ester works well as an additive in diesel fuel. For instance, a study by Sulek *et al.* [36] demonstrated the incorporation of fatty acid methyl ester into diesel fuel produced from rapeseed oil that minimized friction by 20% and increased wear by twice as much as available diesel oil. Bekal and Bhat [37] tested the performance of mineral oil after blending pongamia oil in various ratios. These findings suggest that mixed oils will perform at their best. In addition, Sukjit and Dearn [28] reported that 5% methyl ester made from rapeseed oil reduced wear scars by 40% when combined with commercial diesel oil. In a related study, Durak [38] mixed different volumes of Turkish-formed rapeseed oil with mineral oil and evaluated how well it lubricated high-temperature engine journal bearings. His research demonstrated that rapeseed oil lowers the friction coefficient.

Since 1983, the transesterification byproduct of crude palm oil, known as palm oil methyl ester (POME), which includes an ester functional group, has been suggested as another possible lubricant additive [39]. The addition of POME to commercial base stock oil has significantly improved friction behavior and wear scar diameter under reduced loads and temperatures. Research by Masjuki and Sapuan [40] revealed greater anti-wear properties in engines when compared to 100% Society of Automotive

Engineer (SAE) 40 lubricating oil. They did this by blending POME with commercial SAE 40 oil (95% SAE 40 + 5% POME). Masjuki and Maleque [41] revised this work by employing a four-ball tribometer test to analyze the anti-wear properties of blended pure lubricating oils with 3%, 5%, 7%, and 10% by volume of POME. A similar investigation was performed by Maleque *et al.* [42]. They drew the conclusion that degradation due to wear and friction parameters is lower at lower loads and temperatures. However, POME exhibits large wear scar diameter, corrosive behavior, and pitting formation at higher temperatures and loads. Kalam and Masjuki [43] did lubricant oil analyses for palm oil diesel, regular diesel, and these oils' emulsions. The collection of metal wear debris in lubricant samples was reduced with the use of palm oil-based diesel and its emulsion compared to regular fuel during a 20-h experiment.

8.3.3 Water-Based Nanolubricants

Oil-based lubricants inevitably contaminate the environment when discharged or burned. Water is a superior source of coolant due to its high specific heat capacity, but it can also be utilized as an affordable, clean and green with environmentally friendly. However, the use of water as a lubricant is restricted because of its poor viscosity and corrosive behavior. This drawback can be minimized by dispersing additives in the nanoform in water [44–48]. As a synthesis process for water-based nanolubricants, the nanoscale particles are dispersed in a water base with the help of a surfactant. This step is followed by mechanical agitation and ultrasonication. Water can be utilized as lubricant in hot steel rolling, etc., and can offer less friction, less wear, and good load-carrying capacities when nanoadditives are incorporated. These additives can be metallic oxides [49, 50], metals [51, 52], non-metallic oxides [53, 54], ceramics [55, 56], metal sulfides [57], carbon materials [58, 59] and composites [60, 61]. One of the most widely explored additives is carbon-based materials, which accounts for 35% of the reported work [62]. A pie chart of different additives used over the last decade is shown in Figure 8.4.

Metals, such as Au, Ag, and Cu, have been considered as additives due to their properties [63]. Due to their weak compatibility with base lubricants, surface modification techniques are used [64]. *In situ* modification was used to fabricate Cu NPs, and then dispersed in water-based lubricants. The Cu surface was capped with methoxylpolyethylene-glycol xanthate potassium for dispersion in water-based lubricants in a work by Zhang *et al.* [65]. The output was such that there were no agglomeration, and the Cu nanoparticle size was 2 nm. Among the widely explored metal oxides as additives

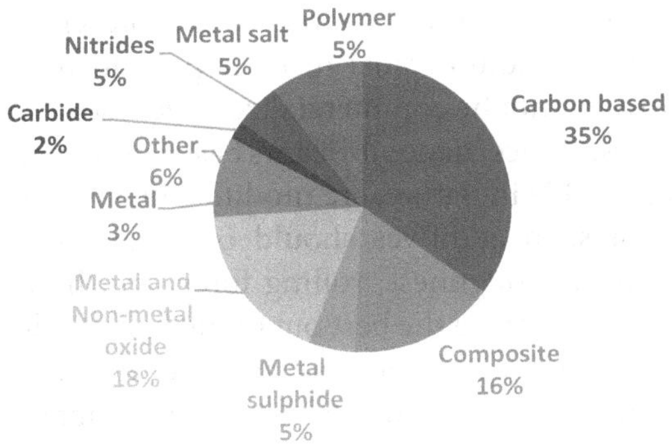

Figure 8.4 A pie chart of different additives used over the last decade.

in water-based lubricants are Al_2O_3, TiO_2, and ZnO [66]. He *et al.* [67] used a MoS_2-Al_2O_3 nanocomposite in water-based lubricants for hot rolling. This reduced the oxide scale thickness and rolling force of the rolled strips. When TiO_2 was used in water lubricants, the thickness of the oxide scale and grain size of rolled steel substantially decreased [50]. The surface properties of hot rolled strips were also modified when water-based lubricants with nano-SiO_2 as additives were used [54]. Different phases of TiO_2 nanoparticles having different sizes have been widely explored and used in hot rolling, tribology, end milling, grinding, and drilling [68]. There has been an increase in the use of nanoadditives for water-based lubricants in various applications. Metal sulfides like CuS, MoS_2, and Ag_2S have superior lubricating properties and are efficient additives. CuS was used as an additive in water-based lubricants by Zhao *et al.*, and it remained uniformly dispersed in water for a minimum of 2 days [69]. MoS_2 as an additive offers superior thermal and chemical stability. Meng *et al.* [70] observed that MoS_2-based water lubricants were stable for 16 h. Functional MoS_2 will offer even better features like better dispersion and more stability [57].

8.4 Conclusion and Outlook

Research on the development of efficient nanolubricants, which are environmentally friendly, has moved forward. Synthesis of nanoadditives-based biolubricants has been one of the most widely conducted research. New materials are still in the developing phases, which will reduce the impact on the environment and help in the green operation of lubricant-based machines. The environmentally friendly nanolubricants should be easily biodegradable, non-toxic, and non-carcinogenic. The synthesis of the same should not degrade the natural resources and avoid creation of harmful decay products. But nanolubricants in their pure form are not efficient enough and mostly cannot be commercialized because of their inferior tribological properties. Hence, nanolubricants based on vegetable oil or water, which are biodegradable in nature, are modified using various additives.

The addition of such additives should optimize the nanolubricants. Effects such as surface roughness, rolling force, and hardness should be accordingly modified to bring the best out of the final biodegradable nanolubricant. When water-based nanolubricants are mixed with appropriate additives, the wettability, dispersion stability, and thermal conductivity are bound to improve. In such lubricants, if optimized, surface roughness and rolling force will reduce along with the oxide scale thickness.

For vegetable oil-based nanolubricants, polar groups and fatty acid chains govern their friendly lubricating properties. To further improve their properties, techniques such as epoxidation, post-esterification, hydrogenation, and transesterification are used. These techniques help overcome their poor thermo-oxidation property. The use of additives is also a way to improve their, but not all additives will improve the properties. Therefore, a selective choice of additives that are not harmful to the surroundings is required. This aspect opens greater avenues for research in the field. Novel compatible additives, which complement and improve the properties of water-based and vegetable oil-based biolubricants have to be a part of the upcoming research in the field. In the case of nanoparticles as additives, the size, shape, and concentration have to be in the right ratio along with their other intrinsic properties. We can possibly conclude that the optimization in size has been explored, and the range of 20–70 nm is suitable. However, research on the influence of shape is an area that warrants further exploration.

References

1. Saravanakumar, N. *et al.*, Development and testing of nano particulate lubricant for worm gear application. *J. Mech. Sci. Technol.*, 33, 4, 1785–1791, 2019.
2. Karthikeyan, K. *et al.*, Thermophysical and wear properties of eco-friendly nano lubricants. *Mater. Today: Proc.*, 39, 285–291, 2021.
3. Rahman, M.M., Development of vegetable oil-based nano-lubricants using montmorillonite clay (MMT) nanoparticles as lubricant additives, The University of Texas Rio Grande Valley, 2021.
4. Darminesh, S.P. *et al.*, Recent development on biodegradable nanolubricant: A review. *Int. Commun. Heat Mass Transfer*, 86, 159–165, 2017.

5. Lee, K. *et al.*, Understanding the role of nanoparticles in nano-oil lubrication. *Tribol. Lett.*, 35, 2, 127–131, 2009.
6. Peng, D. *et al.*, Tribological properties of diamond and SiO_2 nanoparticles added in paraffin. *Tribology Int.*, 42, 6, 911–917, 2009.
7. Spikes, H., Film-forming additives-direct and indirect ways to reduce friction. *Lubr. Sci.*, 14, 2, 147–167, 2002.
8. W.M.L. Suer, Lubrizol Corp, (Patent 2247534).
9. J.H. Cleverley, J.F.M., J.M. Swietlik, Exxon Chemical Corp, (Patent EP 0323087).
10. Martin, J.M. and Ohmae, N., *Nanolubricants*, John Wiley & Sons, England, 2008.
11. Peterson, M.B., Friction and wear investigation of molybdenum disulfide I:Effect of moisture, 1953.
12. Winer, W.O., Molybdenum disulfide as a lubricant: A review of the fundamental knowledge. *Wear*, 10, 6, 422–452, 1967.
13. Mobarak, H. *et al.*, e prospects of biolubricants as alternatives in automotive applications. *Renewable Sustainable Energy Rev.*, 33, 34–43, 2014.
14. Singh, A., Chauhan, P., Mamatha, T., A review on tribological performance of lubricants with nanoparticles additives. *Mater. Today: Proc.*, 25, 586–591, 2020.
15. Uflyand, I.E., Zhinzhilo, V.A., Burlakova, V.E., Metal-containing nanomaterials as lubricant additives: State-of-the-art and future development. *Friction*, 7, 2, 93–116, 2019.
16. Zhao, J. *et al.*, Nanolubricant additives: A review. *Friction*, 9, 5, 891–917, 2021.
17. Gulzar, M. *et al.*, Tribological performance of nanoparticles as lubricating oil additives. *J. Nanopart. Res.*, 18, 8, 1–25, 2016.
18. Kong, L., Sun, J., Bao, Y., Preparation, characterization and tribological mechanism of nanofluids. *RSC Adv.*, 7, 21, 12599–12609, 2017.
19. Lee, C.-G. *et al.*, A study on the tribological characteristics of graphite nano lubricants. *Int. J. Precis. Eng. Manuf.*, 10, 1, 85–90, 2009.
20. Kim, H.-S. *et al.*, Tribological characteristics of paraffin liquid with nanodiamond based on the scuffing life and wear amount. *Wear*, 301, 1-2, 763–767, 2013.
21. Anand, O. and Chhibber, V.K., Vegetable oil derivatives: Environment-friendly lubricants and fuels. *J. Synth. Lubr.*, 23, 2, 91–107, 2006.
22. Boyde, S., Green lubricants. Environmental benefits and impacts of lubrication. *Green Chem.*, 4, 4, 293–307, 2002.
23. Kitamura, N., Biodegradable lubricants. *Jpn. J. Tribol.*, 38, 5, 639–648, 1993.
24. Erhan, S.M. and Kleiman, R., Biodegradation of estolides from monounsaturated fatty acids. *J. Am. Oil Chem. Soc.*, 74, 5, 605–607, 1997.
25. Kassfeldt, E. and Dave, G., Environmentally adapted hydraulic oils. *Wear*, 207, 1-2, 41–45, 1997.

26. Norrby, T., Environmentally adapted lubricants–where are the opportunities? *Ind. Lubr. Tribol.*, 55, 6, 268–274, 2003.
27. Pettersson, A., High-performance base fluids for environmentally adapted lubricants. *Tribol. Int.*, 40, 4, 638–645, 2007.
28. Hsien, W.L.Y., *Towards green lubrication in machining*, Springer, Singapore, 2015.
29. Rudnick, L.R., *Lubricant additives: Chemistry and applications*, CRC Press, USA, 2009.
30. Fox, N. and Stachowiak, G., Vegetable oil-based lubricants—A review of oxidation. *Tribol. Int.*, 40, 7, 1035–1046, 2007.
31. Ruger, C.W., Klinker, E.J., Hammond, E.G., Abilities of some antioxidants to stabilize soybean oil in industrial use conditions. *J. Am. Oil Chem. Soc.*, 79, 7, 733–736, 2002.
32. Joseph, P. and Sharma, D., Improvement of thermooxidative stability of non-edible vegetable oils of Indian origin for biodegradable lubricant application. *Lubr. Sci.*, 22, 4, 149–161, 2010.
33. Pokorny, J., Major factors affecting the autoxidation of lipids, in: *Autoxidation of Unsaturated Lipids*, pp. 141–206, 1987.
34. Syahrullail, S., Kamitani, S., Shakirin, A., Performance of vegetable oil as lubricant in extreme pressure condition. *Proc. Eng.*, 68, 172–177, 2013.
35. Kumar, B.S., Padmanabhan, G., Krishna, P.V., Performance assessment of vegetable oil based cutting fluids with extreme pressure additive in machining. *J. Adv. Res. Mater. Sci.*, 19, 1, 1–13, 2016.
36. Salih, N., Salimon, J., Yousif, E., Synthetic biolubricant basestocks based on environmentally friendly raw materials. *J. King Saud Univ.-Sci.*, 24, 3, 221–226, 2012.
37. Zulkifli, N.W.M. *et al.*, The effect of temperature on tribological properties of chemically modified bio-based lubricant. *Tribol. Trans.*, 57, 3, 408–415, 2014.
38. Sulek, M., Kulczycki, A., Malysa, A., Assessment of lubricity of compositions of fuel oil with biocomponents derived from rape-seed. *Wear*, 268, 1-2, 104–108, 2010.
39. Bekal, S. and Bhat, N., Bio-lubricant as an alternative to mineral oil for a CI engine—An experimental investigation with pongamia oil as a lubricant. *Energy Sources Part A: Recovery Util. Environ. Eff.*, 34, 11, 1016–1026, 2012.
40. Sukjit, E. and Dearn, K., Enhancing the lubricity of an environmentally friendly Swedish diesel fuel MK1. *Wear*, 271, 9-10, 1772–1777, 2011.
41. Maleque, M., Masjuki, H., Sapuan, S., Vegetable-based biodegradable lubricating oil additives. *Ind. Lubr. Tribol.*, 55, 3, 137–143, 2003.
42. Canter, N., Special report: Trends in extreme pressure additives. *Tribol. Lubr. Technol.*, 63, 9, 10, 2007.
43. Masjuki, H. and Sapuan, S., Palm oil methyl esters as lubricant additive in a small diesel engine. *J. Am. Oil Chem. Soc.*, 72, 5, 609–612, 1995.

44. Wen, P. *et al.*, Synergy between covalent organic frameworks and surfactants to promote water-based lubrication and corrosion resistance. *ACS Appl. Nano Mater.*, 3, 2, 1400–1411, 2020.
45. Su, F., Chen, G., Huang, P., Lubricating performances of graphene oxide and onion-like carbon as water-based lubricant additives for smooth and sand-blasted steel discs. *Friction*, 8, 1, 47–57, 2020.
46. Chen, W. *et al.*, Macroscopic friction studies of Alkylglucopyranosides as additives for water-based lubricants. *Lubricants*, 8, 1, 11, 2020.
47. He, A. *et al.*, Tribological characteristics of aqueous graphene oxide, graphitic carbon nitride, and their mixed suspensions. *Tribol. Lett.*, 66, 1, 1–12, 2018.
48. del Río, J.M.L. *et al.*, Improving the tribological performance of a biodegradable lubricant adding graphene nanoplatelets as additives. *J. Mol. Liq.*, 345, 117797, 2022.
49. Pardue, T. *et al.*, A tribological study of γ-Fe_2O_3 nanoparticles in aqueous suspension. *Tribol. Lett.*, 66, 4, 1–13, 2018.
50. Meng, Y. *et al.*, The role of nano-TiO_2 lubricating fluid on the hot rolled surface and metallographic structure of SS41 steel. *Nanomaterials*, 8, 2, 111, 2018.
51. Zhao, J. *et al.*, Synthesis of water-soluble Cu nanoparticles and evaluation of their tribological properties and thermal conductivity as a water-based additive. *Friction*, 7, 3, 246–259, 2019.
52. Song, H. *et al.*, Facile synthesis of core–shell Ag@C nanospheres with improved tribological properties for water-based additives. *New J. Chem.*, 42, 11, 8773–8782, 2018.
53. Ding, M. *et al.*, The excellent anti-wear and friction reduction properties of silica nanoparticles as ceramic water lubrication additives. *Ceram. Int.*, 44, 12, 14901–14906, 2018.
54. Bao, Y., Sun, J., Kong, L., Effects of nano-SiO_2 as water-based lubricant additive on surface qualities of strips after hot rolling. *Tribol. Int.*, 114, 257–263, 2017.
55. Lin, B. *et al.*, Excellent water lubrication additives for silicon nitride to achieve superlubricity under extreme conditions. *Langmuir*, 35, 46, 14861–14869, 2019.
56. He, J. *et al.*, Preliminary investigations on the tribological performance of hexagonal boron nitride nanofluids as lubricant for steel/steel friction pairs. *Surf. Topogr.: Metrol. Prop.*, 7, 1, 015022, 2019.
57. Wang, Y. *et al.*, Friction reduction of water based lubricant with highly dispersed functional MoS2 nanosheets. *Colloids Surf. A: Physicochem. Eng. Asp.*, 562, 321–328, 2019.
58. Ye, X., Songfeng, E., Fan, M., The influences of functionalized carbon nanotubes as lubricating additives: Length and diameter. *Diamond Relat. Mater.*, 100, 107548, 2019.

59. Tang, W. *et al.*, Facile pyrolysis synthesis of ionic liquid capped carbon dots and subsequent application as the water-based lubricant additives. *J. Mater. Sci.*, 54, 2, 1171–1183, 2019.
60. Huang, S. *et al.*, Synergistic tribological performance of a water based lubricant using graphene oxide and alumina hybrid nanoparticles as additives. *Tribol. Int.*, 135, 170–180, 2019.
61. Wu, P. *et al.*, Synergistic tribological behaviors of graphene oxide and nanodiamond as lubricating additives in water. *Tribol. Int.*, 132, 177–184, 2019.
62. Morshed, A., Wu, H., Jiang, Z., A comprehensive review of water-based nanolubricants. *Lubricants*, 9, 9, 89, 2021.
63. Haddad, Z. *et al.*, A review on how the researchers prepare their nanofluids. *Int. J. Therm. Sci.*, 76, 168–189, 2014.
64. Shahnazar, S., Bagheri, S., Abd Hamid, S.B., Enhancing lubricant properties by nanoparticle additives. *Int. J. Hydrogen Energy*, 41, 4, 3153–3170, 2016.
65. Zhang, C. *et al.*, Preparation and tribological properties of surface-capped copper nanoparticle as a water-based lubricant additive. *Tribol. Lett.*, 54, 1, 25–33, 2014.
66. Radice, S. and Mischler, S., Effect of electrochemical and mechanical parameters on the lubrication behaviour of Al2O3 nanoparticles in aqueous suspensions. *Wear*, 261, 9, 1032–1041, 2006.
67. He, J. *et al.*, Superior lubrication performance of MoS_2-Al_2O_3 composite nano-fluid in strips hot rolling. *J. Manuf. Processes*, 57, 312–323, 2020.
68. Dhanola, A. and Garg, H., Dispersion stability and rheology study of canola oil containing TiO_2 nanoadditives for tribological applications: An experimental approach. *Proc. Inst. Mech. Eng., Part J: J. Eng. Tribol.*, 235, 9, 1765–1781, 2021.
69. Zhao, J. *et al.*, A simple preparation of HDA-CuS nanoparticles and their tribological properties as a water-based lubrication additive. *Tribol. Lett.*, 67, 3, 1–11, 2019.
70. Yanan, M. *et al.*, Recycling prospect and sustainable lubrication mechanism of water-based MoS_2 nano-lubricant for steel cold rolling process. *J. Cleaner Prod.*, 277, 123991, 2020.

9

Nanodimensional Metal-/Metal Oxide-Incorporated Vegetable Oil-Based Biodegradable Lubricants: Environmental Benefits, Progress, and Challenges

Pooja Sharma[1], Umar Farooq[2*], Syed Salman Ali[3] and Kaneez Fatima[4]

[1]Department of Pharmacy, BSA College of Engineering and Technology, Mathura, Uttar Pradesh, India
[2]Chemistry Division, Department of Basic Sciences, School of Basic and Applied Sciences, Galgotias University, Greater Noida, India
[3]Lloyd Institute of Management and Technology, Plot No-11, Knowledge Park-II, Greater Noida, UP, India
[4]Department of Pharmacy, Maulana Azad University, Jodhpur, Rajasthan, India

Abstract
Lubricants include substances that can facilitate effortless relative motion between two or more solid bodies by reducing the wear and friction between the interacting bodies. Currently, most of the lubricants are non-aqueous and nonbiodegradable in nature. Therefore, at present, most research developments are focusing on environmentally benign, biodegradable fluids for lubrication and greasing purposes. Based on the European Union Ecolabel, in the fabrication and selection of base materials and final formulized lubricants, biodegradability and environment friendly nature are considered as two of the most important design parameters. Like other fields, biodegradable lubricants are in great demand due to the adverse impact of technological developments on the environment. Among different environmentally benign lubricants explored, vegetable oil-based lubricants highly attract researchers. Their renewable, environment-friendly nature, and marketing advantages have caught the interest of many multinational companies. The benefits of having vegetable oils as lubricants have made a room for their use in drilling fluids, two stroke engines, gears, metalworking, greases, outboard engine

**Corresponding author*: darumer27@gmail.com; umar.farooq@galgotiasuniversity.edu.in

Mohd Yusuf, Lalit Prasad and Shafat Ahmad Khan (eds.) Nanolubricants: Generation and Applications, (229–256) © 2024 Scrivener Publishing LLC

lubricants, as well as hydraulic fluids, mould release oils, and chainsaw oils. To meet the demands of efficient lubrication, the incorporation of different additives, particularly at nanoscale, is being explored. Lubricants with nanoparticles as additives show efficient friction-reducing and anti-wear properties. In this chapter, we will discuss the development of vegetable oil nanodimensional metal/metal oxide-based lubricants for their different lubricating applications. Also, the effect of the addition of nanoparticles in various properties like viscosity, volatility, toxicity, and shear stability of vegetable oil-based biodegradable lubricants will be discussed in detail. In addition, the limitations and progress to be made in the widespread use of biodegradable vegetable oil-based lubricants will be discussed.

***Keywords*:** Nanolubricant, biodegradable, vegetable oil, additives

9.1 Introduction

The continuous exponential growth of the human population along with the development of new technologies and industries have serious repercussions on the good healthy biotic and abiotic components of the ecosystem/environment. The increasing population leads to the rise in energy demand, which leads to consumption of more and more petroleum-based fuels, rise in greenhouse gas emission, and climate change [1–3]. Presently, most of the industries are run by using petroleum-derived products. The damage caused by these petroleum-based products to the environment has forced governments and non-government organizations (NGOs) across the globe to enforce different laws to prevent environmental pollution. Governments all over the world are focusing on the development of target-oriented manufacturing/development of environmentally benign products causing the least harm to the environment. Therefore, to preserve energy resources and the environment for future generations, there is huge interest in the development of have bio-economical energy resources and their associated products. As per the European Union's Sustainable Consumption, Production and Industry Action Plan, a sustainable industrial policy was introduced in 2008 in which the concept of ecolabel was incorporated [4]. The main concept and aim of the EU ecolabel were to promote products with less adverse impact on the environment during their entire lifetime and to spread awareness among consumers about the environmental impacts of products they use. The EU ecolabel will be obtained by a particular product group whose products can reduce negative impacts on the environment compared to other products of that group. Ecolabel has included the product group "lubricants", which comprises of greases, chainsaw oils, engine oils, hydraulic oils, and concrete release oils [5–7].

Lubricants are fluids that are primarily used to reduce friction and wear among different solid movable parts of any machine. Apart from reducing wear and friction, lubricants perform certain other functions like prevention of corrosion, the heat transfer process, the transfer of power, and removal of wear particles [8].

Among different petroleum-based products, the lubricant industry is one of the most important industries based on the incorporation of mineral and synthetic oils during their formulation. Due to its low cost, ease of availability, and diverse viscosity, petroleum-based lubricants, including hydrocarbons (mixture of C20–C30), mineral oils, and many other chemical compounds like aromatic, naphthenic, and paraffinic species are widely used (90%–95%) in different industries [8]. The current worldwide demand for lubricants ranges from 45–50 million tonnes, and out of these lubricants, more than 50% end up in the environment. In addition, the usage of mineral oil-based lubricants is associated with certain drawbacks like the volatilization of components with low molecular weight and compositional differences based on their petroleum source, toxicity, and environmental pollution due to the evolution of oxides of S, N, and C [9, 10]. In the past few decades, general awareness regarding climate change has been observed, and in this perspective, special attention is given to this sector for the development of environment-friendly, biobased, nontoxic biolubricants with efficient lubricating properties [11–14]. Based on this fact, the relevance of the biodegradability of the base oil has been reinforced and has become one of the basic criteria for selection, the design of the base oil and formulation of the finalized product in the form of a lubricant. For the development of lubricants, biodegradability is not the only condition that concerns researchers. The overall lubrication performance of the lubricant is another important parameter [15]. Therefore, for the development of environment-friendly lubricants, biodegradability and performance should harmonize each other to meet both technological and ecological demands. Although biolubricants are expensive compared to mineral oil-based lubricants, they show significant superior properties like high viscosity index, high flash point, less toxicity, high anticorrosion properties, renewable raw material, and good biodegradability in comparison to mineral oil-based lubricants [16–21]. To date, a significant number of biolubricants, including poly-α-olefin with low viscosity, dibasic acid esters, and high oleic vegetable oils, have been developed [22–24].

Besides having unique properties and functionalities, there are certain limitations associated with vegetable oil-based lubricants. These limitations include instability towards oxidation; low resistance towards the degradation process, thus changing their tribological and other properties;

and viscosity change with temperature and hydrolysis [25]. Therefore, to overcome these limitations, different approaches including the incorporation or development of more saturated vegetable oils and incorporation of antioxidants, chemical additives, high oleic acid, and vegetable oils have been investigated. In the past decades, the addition of additives, particularly at nanoscale, has attracted significant attention. The addition of nanoadditives can significantly improve friction-reducing and anti-wear properties of vegetable oil-based biolubricants [26–30]. In this chapter, we will discuss the different basic characteristics of lubrication, the progress in and the properties and advantages of vegetable oil-based nano-biolubricants.

9.2 Concept of Lubrication and Characteristics of a Lubricant

Whether mineral oil- or vegetable oil-based lubricants, there are some important basic concepts and terms used to discuss the properties and characteristics of any lubricant or the lubrication process. These concepts or terms include friction, wear, and lubrication. In this section, we briefly discuss these concepts and their applications, in the process explaining the behavior of any lubricant.

9.2.1 Friction

When two surfaces in contact slide and roll against each other, there is a certain force of resistance against the tangential motion of these sliding and rolling surfaces. This resistance force is called friction. This resistance or friction arises from the microscopic contact between the surfaces. In most industrial applications like machines possessing gears and bearings, the frictional forces result in conversion of mechanical energy into heat, resulting in the increase in temperature between the sliding or rolling surfaces. These factors could change the properties and tribological behavior of the system, which might result in the failure of the mechanical system. The friction between two bodies is expressed as coefficient of friction (m), which is defined as force per unit load over the sliding bodies (Equation 9.1).

$$M = F/W \tag{9.1}$$

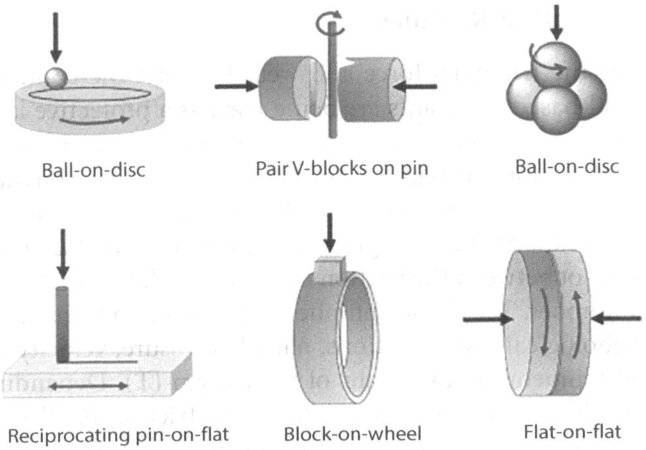

Figure 9.1 Different configurations used in commercial tribometers with normal load pointed by the arrow (black). Reprinted with permission from (ACS Sustainable Chem. Eng. 2021, 9, 1459–1476). 2021 American Chemical Society.

Commercial tribometers are employed to measure tribological properties like the coefficient of friction and wear between two surfaces. Different configurations of tribometers, as shown in Figure 9.1, are employed for the replications of real systems.

9.2.2 Wear

The friction between two sliding or rolling surfaces is responsible for irreversible wear, surface damage, and non-renewable energy losses if in the form of heat. This progressive loss of a surface material forms because of a frictional process known as wear. Depending on the type of observed surface damage on worn surfaces, wear can be classified as adhesion, abrasion, delamination, surface fatigue, etc. The description of wear can be given in three steps [31]: (i) when particles are detached from the body surface by any of the processes like adhesion and abrasion; (ii) a reduction of interaction between two surfaces due to entrapment and circulation of small particles between these bodies, resulting in the formation of a powder bed between them; and (iii) finally, the expulsion of these powder beds takes place, and two bodies again show increased interaction, thus resulting in the formation of wear, and this cycle begins again.

9.2.3 Lubrication Regimes

To minimize the frictional force and wear between the contact surfaces in motion, oil-based lubricants are being used as a protective layer. Based on a normal load between the two moving surfaces in contact, different lubrication conditions or regimes (mixed, boundary, hydrodynamic, and elastohydrodynamics) arise. Different lubrication regimes are explained with the help of the Stribeck diagram as shown in Figure 9.2. The Stribeck diagram is a plot between the friction coefficient of lubricated surfaces and the Hersey number. The Hersey number is the function (dimensionless) of the three tribological test conditions, namely pressure, velocity of sliding/rolling, and temperature by means of viscosity n (T). Depending on the value of the Hersey number, the coefficient of friction (COF), thickness, and lubrication properties of any lubricant can be characterized. At a high Hersey number, a viscous dragging force of a lubricant results in the linear increase in COF. However, with an increase in load and decrease in viscosity and sliding load, the Hersey number (HN) decreases. At a hydrodynamic lubrication, the viscosity of lubrication layer is thick enough (> 1 um) to avoid the contact between two surfaces, thus showing low friction and surface wear [32]. With the decrease in HN, the thickness of the lubrication is further reduced and results in conditions like elastohydrodynamic lubrication (where piezo-viscosity and elastic deformation become significant) and mixed lubrication regime (at which HN further reduces

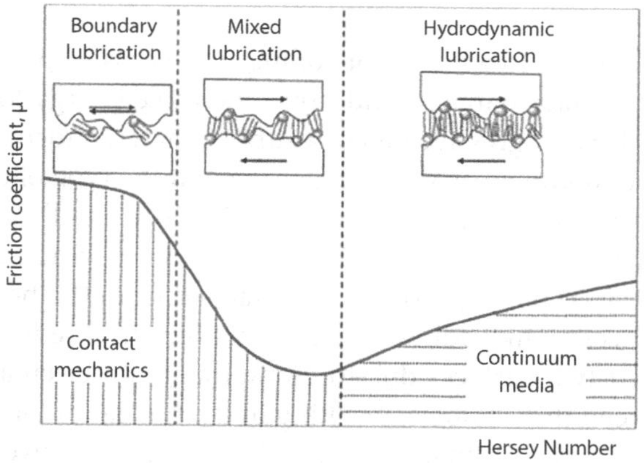

Figure 9.2 An illustration of different lubricating regimes and its relation to the coefficient of friction, speed, and lubricant viscosity. Reprinted with permission from (ACS Sustainable Chem. Eng. 2021, 9, 1459–1476). 2021 American Chemical Society.

due to reduction in the thickness of lubrication film resulting to the start of surface contact). Then, with further reduction in the HN boundary, the lubrication regime starts, causing friction and wear to become more severe.

9.2.4 Characteristics of a Lubricant

To perform the basic role of lubrication, lubricants should have specific chemical and physical characteristics. The fundamental and basic requirement of any lubricant is that it should be in a liquid form over a broad spectrum of temperature. Practically, the liquid range usability of a lubricant is limited by pour point and the flash point at low temperatures and at high temperatures, respectively. For a suitable lubricant, the pour point should be low enough that it can be easily pumped in an equipment, which starts at low temperature, while the flash point for a lubricant should be high for the minimum volatilization and safe operation at high temperature. The viscosity of a lubricant is among one of the most important properties, which affects the efficient application of a lubricant. Figure 9.2 shows the dependence of the coefficient of friction on the coefficient of viscosity in any lubricated system. Two different regions are present in Figure 9.2, one which corresponds to the hydrodynamic lubrication and the other corresponding to the boundary lubrication. A systematic sufficient hydrodynamic pressure is produced at high viscosity, high speed or low load, which fully separates the solid surfaces by a thin film of a lubricant, and this condition is known as a hydrodynamic lubrication. While at low viscosities and speed or high loads, the generation of hydrodynamic pressure is insufficient to separate the solid surfaces, thus the opposite solid surfaces come in contact, and this type of condition corresponds to boundary lubrication. For minimum friction and wear, equipment are devised in such a way that they possibly have hydrodynamic lubrication. The viscosity of a lubricant for optimum operation depends on the load (to be supported) and relative velocities between the surfaces. Since in different machines or in different parts of the same machine, the velocities, as well as the load they are supporting, varies considerably, it is therefore imperative to have lubricants with a wide range of viscosities.

Some important characteristics of lubricants are the temperature, pressure, and shear rate (in some cases) dependence of their viscosity. For a suitable lubricant, the lubrication properties or viscosity should not change over an operational temperature range. The viscosity of a particular lubricant varies significantly with temperature and is usually expressed in terms of their viscosity index. A higher value of viscosity index represents higher stability or lower change in viscosity over an operational temperature range. Similarly, pressure inside the machines working under lubrication conditions is an important parameter influencing the viscosity of the lubricants. The hydrodynamic pressure in some lubricated mechanical parts exceed 1 GPa, under which the viscosity of the lubricants may increase by several orders compared to the viscosity of the same lubricating fluid under ambient conditions. Under such high-pressure lubrication conditions, many lubricants are transformed into solids. However, in many cases, such viscosity increase is beneficial to maintain the liquid film of the lubricants between contacts, thus preventing the failure of machine components. In many machines, such increase in viscosity may be detrimental and may result in the failure of contacts. Therefore, for a proper practical application, the lubricants must maintain its physical characteristics for a longer time due to operational high temperature and oxygen conditions. Hence, for better performance, the lubricants should have high thermal degradation stability and resistance to auto-oxidation.

The inertness of lubricants is another important characteristic, which determines its usage and application under different environmental conditions. For an ideal application, the lubricants should be highly nonreactive with different mechanical parts of a machine.

9.3 Vegetable Oil-Based Biolubricants

Depending on the properties, source, and mode of synthesis, lubricants can be classified as mineral oils, synthetic, and biodegradable. These base oil lubricants are further subclassified as shown in Figure 9.3. Herein, our focus is to discuss biodegradable vegetable oil-based lubricants and the advances made to improve the lubrication characteristics of these vegetable lubricants. Vegetable oil-based lubricants are generally made up of trigylceraldehydes, and are extracted by using a solvent extraction method. During the extraction process, the plant parts are dissolved in solvents like water. The extraction of vegetable oils is also done with the help of a pressure application in which seeds or other parts are pressed under high pressure to extract oil from these plant parts. Based on their biological nontoxic nature, high viscosity, biodegradability, and low viscosity, these vegetable base oils can be employed as important alternatives for mineral oil-based lubricants. Based on the dominating fatty acid present, different types of vegetables are present like oleic linoleic oils (containing palm oil, cotton seed oil, sunflower oil, and peanut oil) having a high percentage of unsaturated fatty acids and ricinolein acid (castor oil containing 89% of ricinolein acid) [33–36].

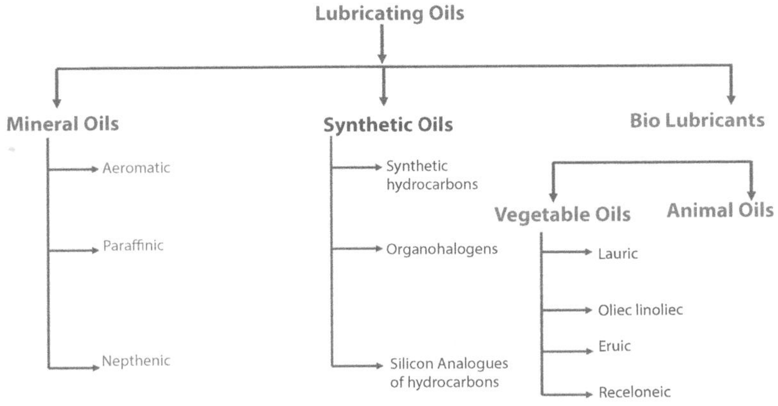

Figure 9.3 Types of base oils for lubricant formulation.

Vegetable oil-based (VO) lubricants are generally liquids at room temperature and find application in all the three regimes and are also more effective in boundary regime. At boundary lubrication regime, strong interactions are developed between the lubricated surface and the VO lubricants due to fatty acid (FA) composition, thus resulting in the formation of a densely packed monolayer, which provides ultralow friction between the surfaces. The formation of this thick, closely packed monolayer between the sliding surface due to the FA present in VO lubricants is a very complex process, which involves different types of interactions (physisorption, chemisorption or tribological reactions) between the sliding/metal surface and the fatty acids. In a physical absorption process, interactions may arise because of weak intermolecular forces like Van der Waals and hydrogen bonding for which the presence of any specific functional group is not required, and this type of interaction can take place in all FAs. However, in case of chemical absorption, there is a sharing of valance electrons between vegetable oil and the lubricated metal surface, and this process is usually irreversible and may sometimes be partially irreversible. In the case of tribological mechanisms, a chemical reaction occurs between a VO lubricant and any other material (like O, S, metal, and H_2O) in the zone of friction. The ability of FAs containing VO to lubricate materials having industrial application in the boundary regime has attracted many researchers, and attempts were made to explain the behavior of the absorption of these lubricants on materials of interest. From these investigations, it was observed that different factors affect the protective tribological properties of VOs under a boundary lubrication regime. These factors include

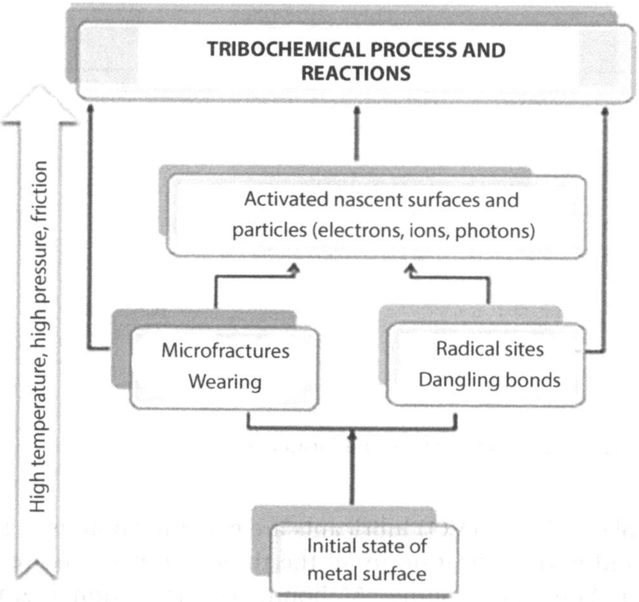

Figure 9.4 The triboemission/process taking place along with the physical and chemical changes during friction. Reprinted with permission from (ACS Sustainable Chem. Eng. 2021, 9, 1459–1476). 2021 American Chemical Society.

the nature of sliding substrate materials, the conformation and unsaturation of fatty acids, the chain length of fatty acids composing the VOs and triboemission (emission of charged particles, valance electrons, photons and lattice components, etc.), as well as tribological chemical reactions under boundary lubrication conditions [37–42]. The process of triboemission due to the changes in surface under friction is illustrated in Figure 9.4.

9.3.1 Limitations of Vegetable Oils (VOs) as Lubricants

The usage of VOs as an efficient lubricant is limited by various factors as discussed below:

9.3.1.1 Auto-Oxidation

As discussed above, VOs offer great potential for the development of important environmentally benign lubricants. The properties and composition of these VOs depend on their original source and are primarily constituted of triacylglycerols (TAGs) (98%). Other constituents of VOs include free fatty acids (FFA), monoacylglycerols (MAGs), diacylglycerols (DAGs), sterols, phosphatides, sterols, tocopherols, and fatty alcohols. The physical and chemical degradation of the biodegradable VOs strongly depend on their composition. Due to the presence of bis-allylic and methylene hydrogens with low bond dissociation energies, the FA and FFA in TAGs show the ability of hydrogen abstraction (Figure 9.5a). Also, the FAs present in most of the VOs are present in cis conformation with a content of trans conformation of less than 4%. Based on the experimental evidence, it has been observed that trans confirmation is more stable as compared to cis conformation. For example, in the case of linoleic acid, the bond dissociation of cis and trans conformation is 328.0 and 334.8 kJ mol^{-1}, which show the preferential oxidation of VOs at cis double-bond sites. Also, the presence of different polar groups in TAGs makes them prone to the attack of oxygen, humidity, physical, and chemical agents as shown in Figure 9.5b.

The main weakness of the presence of unsaturated double bonds is that they act as active sites for different reactions, particularly for self-oxidation in FFAs and TAGs. The mechanism of the auto-oxidation process of FAs involves a free radical chain reaction and comprises of three main steps (initiation, propagation, and chain termination). The auto-oxidation process is initiated because of the presence of thermodynamically favorable reactive oxygen species, which causes the atomic hydrogen abstraction forms TAGs and FAs, as shown in Figure 9.6a. This abstraction of hydrogen preferably occurs at the sites having low bond dissociation energies, i.e.,

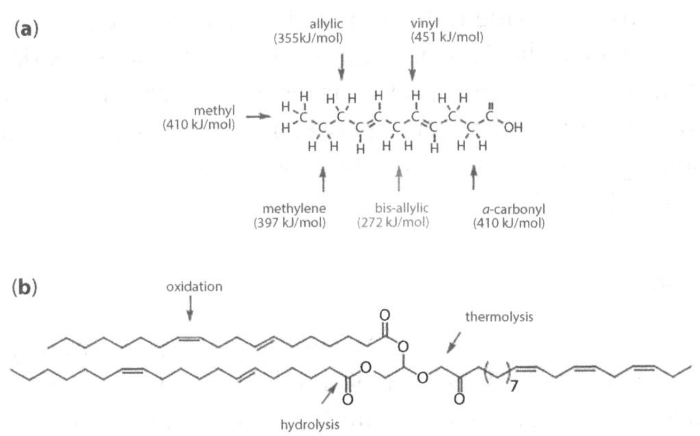

Figure 9.5 (a) Bond dissociation energies of C–H bond and structure and (b) attackable sites (physical and chemical) in TAGs. Reprinted with permission from (ACS Sustainable Chem. Eng. 2021, 9, 1459–1476). 2021 American Chemical Society.

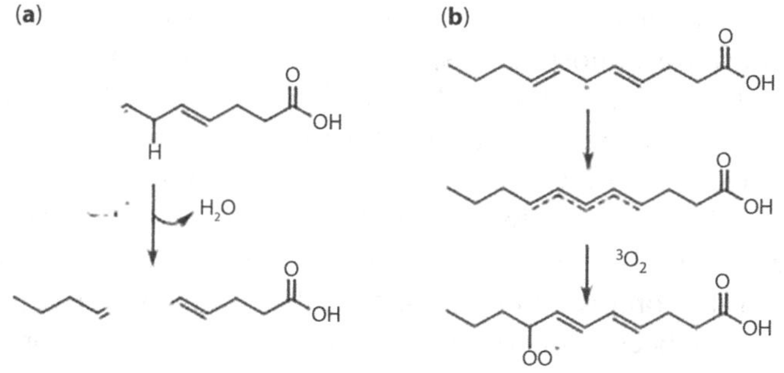

Figure 9.6 Reaction steps involved in auto-oxidation: (a) initiation process of oxidation of lipids and (b) formation of conjugated dienes and hydroperoxide radicals. Reprinted with permission from (ACS Sustainable Chem. Eng. 2021, 9, 1459–1476). 2021 American Chemical Society.

from allylic hydrogens, as shown in Figure 9.6. Therefore, based on bond dissociation energies, the reactivity order towards hydrogen abstraction is as doubly allylic, H ≫ singly allylic H > H α-carbonyl > H methylene, farther down the acyl chain. After the abstraction of hydrogen atoms from TAGs and FAs, a carbon-cantered lipid radical is left, which is stabilized because of the process of molecular resonance rearrangements as shown in Figure 9.6b. Then, the auto-oxidation process is propagated because of the formation of a new alkyl peroxyl radical, which is formed because of the reaction between radicals formed in the initiation step and $3O_2$. The peroxyl radical formed removes hydrogen from the other lipid molecule, resulting in the formation of hydroxy peroxide radicals and a new free radical. The hydroxy peroxide radical may dissociate further, resulting in the formation of more radicals, which can propagate the auto-oxidation reaction of lubricants. This chain reaction is terminated either by the reduction of FAs substrate or by the reaction between alkyl and peroxyl radical, thus either resulting in the formation of nonvolatile compounds or nonradical volatile species [43, 44].

9.3.1.2 Photo-Oxidation of VOs

In addition to auto-oxidation of VOs, another oxidation pathway may take place in the presence of UV light. VO-based lubricants may contain chlorophyll, which acts as an antioxidant under dark conditions. However, upon absorption of UV light, chlorophylls get excited to singlet state, and due to non-radiative intersystem crossing, triplet states are populated. These triplet state chlorophylls can easily react with triplet oxygen, resulting in the formation of singlet oxygen species. The formation of singlet oxygen takes place because of an energy transfer process. As a result, the excited chlorophyll molecules come to its ground state. The obtained singlet oxygen molecule can now easily react with the double bonds (electron-rich centers) present in FAs, resulting in the formation of nonconjugated and conjugated hydroperoxides with no formation of alkyl radicals. Therefore, it is worth mentioning that oxygen molecules play different roles under different conditions, like under the auto-oxidation process, oxygen is in triplet state and does not react directly with electron-rich double bonds while under the photo-oxidation process, oxygen is present in a singlet state and reacts readily with double bonds present in VOs [25].

9.3.1.3 Thermal Oxidation of Vegetable Oils

Under frictional force, mechanical energy is transformed into thermal energy. Therefore, the thermal stability of the lubricants is one of the most important characteristics for efficient lubrication. Hence, most pure VOs cannot be used in high performance applications due to low thermal stress and low operational temperature (70°C). At low temperatures, VOs may show poor flow behavior. The temperature-dependent oxidation of vegetable oil-based lubricants has been vigorously analyzed, but still, the mechanism is not clear as the process involves complex reactions. Due to high temperature, a bond cleavage may take place in TAGs, which could result in the formation of a variety of fragments including lactone, hydrocarbons, and free radicals, which may result in the initiation of the oxidation process. In addition, high temperatures may result in the loss of viscosity, thus giving rise to insoluble polymeric mixture (sludge) or varnish, etc. Other factors that contribute to the Achilles heel of VOs as lubricating agents include degradation due to metal ions and moisture contamination [25].

9.4 Nanolubricants

As discussed above, vegetable oils show the property of degradability, which has both advantages (environmental benign) and disadvantages (poor oxidative stability limiting the direct application in lubrication process). Therefore, to overcome these shortcomings of VOs, lubricants' different modification, physical or chemical, can be done. Among the different modifications for effective positive lubrication, the development and addition of different additives in VOs have grown rapidly in the past decade. The addition of additives, particularly in nanodimensions, has attracted immense interest from researchers due to the unique properties of materials at nanoscale. The incorporation of nanodimensional additives in VOs, hereafter referred to as nanolubricants, offers a wide range of improved lubrication properties like excellent anti-wear and reduced friction. The reduction of friction and excellent anti-wear properties of nanolubricants are attributed to different effects like the mending effect, rolling effect, polishing effect, and formation of protective films [45]. Figure 9.7 represents the mechanism of different effects for improving the tribological properties of VOs by employing nanoparticles.

9.4.1 Mending Mechanism

The mending effect of lubrication using nanoparticles was explained by Liu *et al.* Under this mechanism of reduction of friction and wear, it is believed that the nanodimensional additives are deposited on the frictional surfaces and compensate for the loss of mass, and this effect of deposition is called the mending effect. The mending effect is usually observed under extreme conditions like high load and temperature. At high temperatures, the suspended nanoadditives in nanolubricants show agglomeration and form clusters. Even after agglomeration, these nanoadditives maintain their spherical shape and get deposited on the friction surfaces or fill the surface groves, thus resulting in the mending effect in nanolubricants. The mending mechanism of nanolubrication is depicted in Figure 9.7a.

9.4.2 Rolling/Ball-Bearing Mechanism

Figure 9.7b represents the mechanism involved in the rolling effect of nanolubricants. In this mechanism, the nanoparticles suspended in nanolubricants act as a ball bearing and offer lubrication due to the rolling effect of nanoparticles between the two sliding surfaces.

9.4.3 Formation of Protective Films

Another important mechanism, which is involved in the lubrication properties of nanolubricants, is the formation of thin protective films as shown in Figure 9.7c. In this mechanism, an amorphous layer is formed by nanoparticles over the sliding or rolling surfaces. Due to the formation of a protective layer between the two sliding surfaces, the real contact area is decreased, thus resulting in the lowering of friction between the surfaces. The reason for the formation of thin protective films is the reaction between the nanoparticles and the metal surfaces. Hu *et al.* have demonstrated the thin film formation mechanism of magnesium borate added in 500 SN base oil.

9.4.4 Polishing

Figure 9.7d represents another important mechanism involved in the enhanced lubrication properties of nanolubricants. In this mechanism, the surface of the rubbing surfaces is being polished because of the presence of nanoparticles in the nanolubricants. Different nanoparticles have been used as additives in base oils to improve the lubrication properties. For example, Xu *et al.* have employed diamond particles as additives in a paraffin oil lubricant, and compared to the pristine paraffin oil, the diamond particle-incorporated and modified paraffin-based lubricant showed enhanced lubrication properties due to the smoothening of the rubbing surfaces at the onset of tribotests.

9.4.5 Types of Nanoadditives

Based on composition, different types of nanoadditives have been employed to improve the lubrication properties of vegetable oil-based biolubricants. These nanoadditives include metal nanostructures, metal oxides, metal sulfides, nanocomposites, carbon nanostructures, and rare earth-based nanoparticles. A large number of metal and metal oxide families have been already reported and analyzed for different applications [46–51]. Here in this section, we discuss the use of metals and metal oxide-based nanoadditives as important agents for the improved lubrication of nanolubricants.

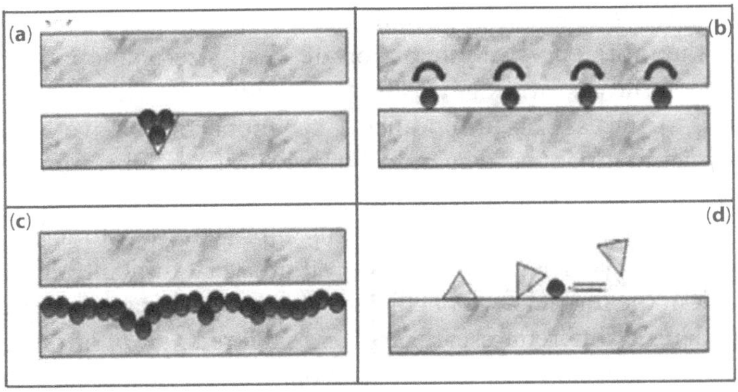

Figure 9.7 Different mechanisms involved in the nanolubrication process.

9.4.6 Vegetable Oil Metal/Metal Oxide-Based Nanolubricants

Numerous metal nanoparticles, including Cu, Ni, Zn, Ag, and Au, have been employed as important additives to improve lubrication properties and stability towards the oxidation process of vegetable oil-based nanolubricants. Among the different metal nanoadditives, Cu nanoparticles have attracted immense interest for nanolubrication application. Zhou *et al.* have developed surface-modified Cu nanoparticles by employing dialkyldithiophosphate (DDP) as a surface modifier [52]. The different molar ratios of Cu:DDP (1:1, 2:1, 4:1, 5:1, and 20:1) were prepared using a constant amount of a Cu precursor and reducing agent and varying concentrations of DDP. The obtained modified and nonmodified Cu nanoparticles were employed as additives in liquid paraffin oil-based lubricant. The anti-wear behavior of DDP-modified Cu nanoparticles additives in paraffin oil was investigated using four ball machines with a ball diameter of 12.7 mm. During synthesis, different sized DDP-modified Cu nanoparticles (2, 5, and 12 nm) were obtained. The unmodified Cu nanoparticles had an average particle size of 40 nm. From the anti-wear analysis, the effective enhanced anti-wear properties were observed in 2 nm DDP-modified Cu nanoparticles added in the paraffin oil lubricant. Similarly, the Cu nanoparticles with 5-nm size show improved lubrication properties. It was observed that the modified Cu nanoparticles with size equal to 12 nm decrease the lubrication properties of liquid paraffin and the nonmodified Cu additives result in the instant seizure of lubrication properties of paraffin oil under similar tribological conditions. The improved lubrication properties of 2 and 5 nm DDP Cu nanoparticles were attributed to the smaller size effect and effective surface interface interaction. As the size of nanoparticles decreases, the melting point also decreases, thus resulting in better ductility, which improves the anti-wear properties of small-sized Cu nanoparticles. Another important factor responsible for effective lubrication is the more effective interaction of small-sized nanoparticles with the sliding surfaces, forming an effective protective surface over the two interacting surfaces. The unmodified Cu nanoparticles have large sizes; thus they act as an abrasive agent, which reduces the lubrication properties of the base oil. Traditionally, for the fabrication of nanolubricants, a two-step route is being employed. In this approach, the storage and dispersion of nanoparticles into the base oil are very difficult and act as a bottleneck in the formulation of efficient bio-nanolubricants. To overcome these challenges, different researchers have explored new *in situ* single-step routes for the development of vegetable oil/mineral oil-based nanolubricants. For cost reduction and industrial applications of nanolubricants,

Zhang *et al.* have developed an *in situ* novel method for the formulation of Ni nanoparticles-based PAO6 [53]. Using this approach, Wang *et al.* have developed *in situ* Cu nanoparticles using a microemulsion of ionic liquids in vegetable oil [54]. The developed Cu-based nanolubricant was used for efficient nanolubrication properties. Water/oil microemulsion acts as an effective nanoreactor for the formulation of nano-Schiff base and Schiff base Cu complexes. However, for the collection of nanoparticles and oil from these microemulsions, de-emulsification is required. Additionally, for the analysis of tribological properties and performance on nanolubricants, the dispersed nanoparticles should be diluted by lubricant base oil. From the available literature, it was observed that the approaches used for the synthesis of Cu nanoparticle-based nanolubricants were not green. However, Wang *et al.* have reported the *in situ* synthesis of Cu nanoparticles using vegetable oil/ionic liquid-based microemulsion. The *in situ* developed Cu-based nanolubricants were directly used for the nanolubrication properties. By employing this method, a highly stable Cu nanoparticle dispersion was prepared in a nonaqueous microemulsion by the reduction process. Reduction was carried out by using a nontoxic reducing

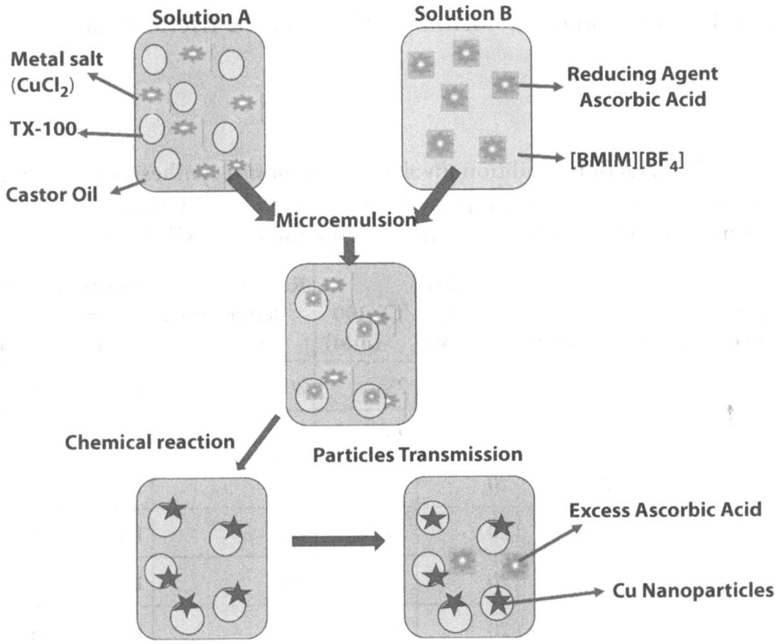

Figure 9.8 Reaction steps involved in *in situ* synthesis of Cu nanoparticles in vegetable oil ionic liquid microemulsion.

agent in the form of ascorbic acid, which also prevents the oxidation of synthesized nanoparticles. Figure 9.8 shows the typical approach used for the *in situ* synthesis of Cu nanoparticles using vegetable oil/ionic liquid microemulsion.

Different reaction conditions, as tabulated in Table 9.1, were employed for the synthesis of different samples. The tribological results obtained for samples S1 and S2, as well as the blank sample, depicted that the blank sample showed fluctuated tribological/frictional coefficient results with high mean value. However, with the incorporation of Cu nanoparticles during *in situ* synthesis, the frictional coefficient varies smoothly over a period. Also, the mean value showed a considerable decrease compared to the blank sample. It was also observed that the wt% value of nanoadditives has a significant effect on the reduction of its frictional coefficient as 0.32 wt% showed a small reduction in frictional coefficient, as compared to the 0.16 wt% of Cu nanoparticles. The wear results obtained from different samples including blank lubricants were investigated. From the wear results, it was observed that lubricants containing Cu nanoparticles show enhanced anti-wear properties compared to the blank lubricant. The observed enhanced properties of nanolubricants were attributed to the fact that due to their small size, the dispersed nanoparticles can penetrate contact surfaces and may be deposited on them, forming a film over the frictional surfaces.

Table 9.1 The reaction condition involved in the *in situ* synthesis of different Cu-based nanolubricants. AA represents ascorbic acid and 1-butyl-3-methylimidazolium tetra-fluoroborate is represented as [BMIM][BF4].

Sample name	AA (mM)	$CuCl_2$ (mM)	[BMIM] [BF_4] (mM)	TX-100 (mM)	Reaction temperature (°C)	Reaction time (h)	Castor oil (mM)
Blank	0	0	20	25	80	1	80
S1	12	4	20	25	80	1	80
S2	6	2	20	25	80	1	80
S3	6	2	20	25	80	1.5	80
S4	12	4	20	25	80	1	80
S5	12	4	20	25	80	1	80

The tribological properties of Pongamia oil (PO) in the presence of graphene nanoplates as additives were investigated by Jason *et al.* [55]. PO is an inedible oil and can therefore be employed for various industrial applications without disrupting the food supply. In this study, a comparative analysis of the effect of graphene nanoplatelets on tribological properties of PO and a mineral oil SAE 15WO (MO) base oil was carried out. Scanning electron microscopic studies were carried out to investigate the worn surfaces after carrying out the tribotesting. During the preparation of graphene platelets-based nanolubricants, 0.01, 0.02, 0.05, 0.1, and 0.2 wt% of graphene were employed. The dispersion of graphene nanoplatelets in base oils was carried out by using a hot plate with continuous magnetic stirring. A four ball tester, as shown in Figure 9.9, was employed to investigate the tribological properties of PO graphene-based nanolubricants.

During the testing process, MO with and without graphene additives were used as a reference to investigate the influence of a PO and PO/graphene nanolubricant on the wear behavior and friction between the contact surfaces. The variation of coefficient of friction (COF) with time for pure and modified MO and PO with different wt% values of graphene platelets is shown in Figure 9.10. From the obtained results, it was observed that for MO series initially, the COF increases up to 400 s, shows the downward trend for 300–600 s, and stays linear afterwards. For the MO sample with 0.2 wt%, graphene nanoplates spike in COF was observed at 600 s,

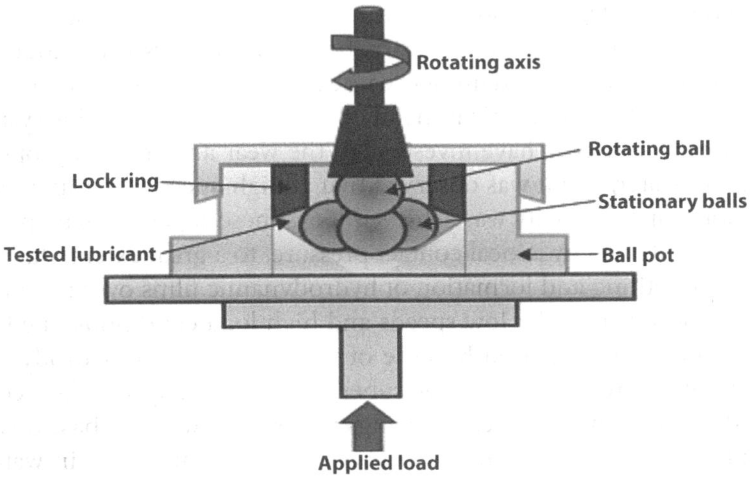

Figure 9.9 Four ball tester for analysis of lubrication properties.

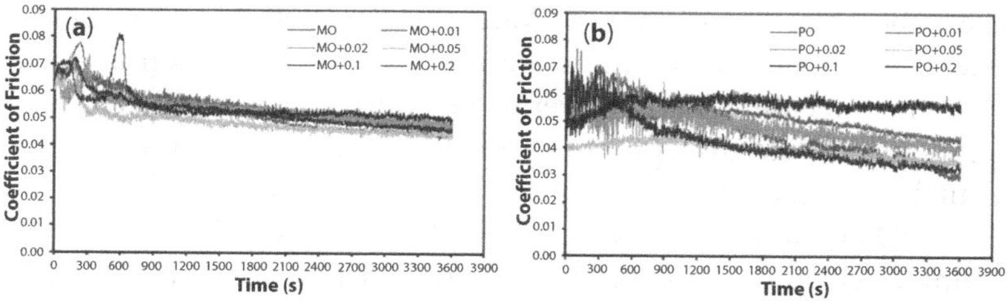

Figure 9.10 The variation of coefficient of friction MO and PO samples modified with different wt% values of graphene nanosheets.

which can be attributed to the decrease in the suspension stability of nanoparticles at higher concentration. Among all the prepared samples, MO + 0.05 shows better anti-friction performance. Figure 9.10b shows the variation of COF of a PO sample containing different wt% of graphene nanoadditives. From the obtained results, it has been observed that pure PO and PO with 0.1 wt% incorporated graphene show an initial increase in COF up to 300 s, then it decreases linearly with time as PO with 0.05 wt% graphene shows stable anti-friction performance throughout the test. From the results, it was observed that there is a certain optimum wt% of graphene, which shows low COF while a further increase in the amount of nanoadditives results in an increase in friction between the contact surfaces, which is attributed to the fact that at higher concentration of nanoparticles, agglomeration takes place, which decreases the viscosity of the nanolubricants, therefore, increasing the friction between surfaces.

The lubrication of diamond-like carbon and ceramics with biodegradable material have attracted immense research interest for sustainable alternatives of non-biodegradable mineral oils lubricating the metallic systems. Tomizawa and Fischer have investigated the wear and friction properties of Si_3N_4 in water, and it was observed that at high-speed water lubrication condition, super lubricity was achieved [56]. These super lubricity properties were explained by a local contact pressure, to a great extent, due to its surface smoothing and formation of hydrodynamic films over the surface of Si_3N_4. However, under low speeds and high load conditions, the lubrication properties decreased because of the breakdown of hydrodynamic lubrication. Therefore, to improve super lubricity under boundary conditions, different nanoadditives are being employed in water or base oils. For example, the addition of nanoparticles of silica was dispersed in water to form a silica gel film, which is homogenously deposited over the Si_3N_4 surface. Another important approach to improve the lubrication properties of

Si_3N_4 is to employ a mixture of water and glycerol under acidic conditions. Luo *et al.* have investigated the lubrication properties of Si_3N_3/glass system and super lubricity was observed using different acid solutions along with glycerol [57]. A two-step mechanism was employed to explain the super-low friction. At pH < 1.5, the protonation of silica under acidic conditions results in the formation of tribolayers, and subsequently the formation of hydrogen bonds between glycerol water system and positively charged silica takes place. Further, the formation of a hydrate layer over glycerol layers takes place, which clearly demonstrates the dependence of lubrication properties of the acidic solution pH. Since, in modern tribological technology, most ceramics contain excess steel, which is also in contact with the lubricating base oil, there are more chances of tribochemical reactions with acidic lubricants, which could result in machine failure. Therefore, to address these issues, nonacidic water-based lubricants are being preferred and investigated for super lubricity. Recently, Long *et al.* have developed an *in situ* approach for the development of a nanolayer of graphene nitride (GN) over Si_3N_4 lubricated with glycerol for super lubricity applications [58]. Super lubricity is a condition where the coefficient of friction is very low (< 0.01). Several earlier reports have reported very low friction of silicon nitride lubricated with water at very high speeds. The ultralow friction was owed to the hydrodynamic water films that formed on silicon hydrophilic surface layers. In this report, boundary lubrication was conducted at low sliding speeds, accompanied by atomistic simulations. An alternative mechanism for the super lubricity of Si_3N_4 lubricated with glycerol was proposed. Various analytical techniques, such as X-ray photoelectron spectroscopy and time-of-flight secondary ion mass spectroscopy, were employed at the wear points, both inside and outside of the wear. The used lubricant was analyzed using a high resolution mass spectroscopy,

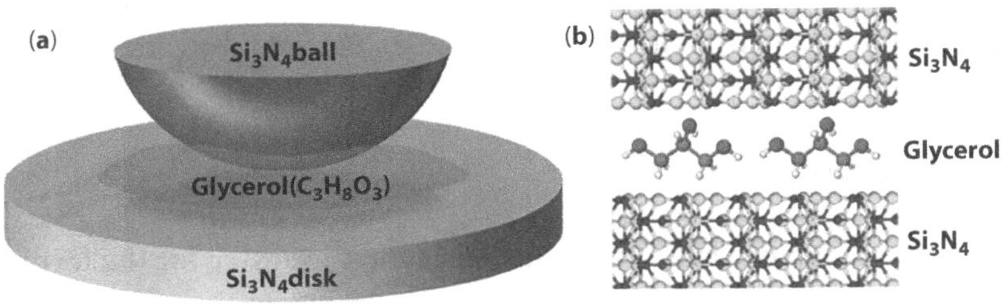

Figure 9.11 Illustrations of (a) an experiment used for the *in situ* synthesis of graphene nitride and (b) computational studies of a system.

and it was observed that the formation of a subnanometer, a thick layer of graphene nitride, takes place over the top of the silicon nitride. From the atomistic simulation results, it was observed that the tribochemical decomposition of glycerol takes place and reacts with nitrogen atoms present in C_3N_2. The formation of 2D graphene nitrides is promoted by shearing, which results in the passivation of ceramic surfaces, thus inducing super lubricity or superlow friction under boundary lubrication conditions. The *in situ* formation of graphitic nitrides is depicted in Figure 9.11.

9.5 Challenges for Sustainable Bio-Nanolubrication

With the development of biolubricants as an alternative for mineral oil-based lubricants, time-to-time modifications are being done to enhance the lubrication properties of vegetable oil lubricants. These modifications have proven to be beneficial in the investigation of lubrication mechanisms and behavior of lubricants, in general, and biolubricants, in particular, under different operating conditions. As discussed, the development of nanolubricants from vegetable oils has been envisioned as an important modification due to which lubrication properties have been enhanced to a great extent. With development of these nanolubricants, there are still different challenges that cause a bottleneck in the development of a versatile lubricant. Among the different challenges, the harmonization of nanoparticles into vegetable oil is a great challenge. As of now, only a few nanoparticles have been successfully incorporated into vegetable oil due to their inability to remain stable and well-dispersed for a longer period. Also, the development of nanoparticles having properties that can enhance the lubrication properties of base oil is also challenging as not every nanoparticle can has the ability to improve its lubrication property. As discussed above, the *in situ* development of nanoparticles for the formulation of efficient nanolubricants is another challenge, which needs further investigation to understand the mechanisms of its formation and enhanced lubrication.

9.6 Conclusion

Currently, most base oils used in the lubrication industry for the formulation of lubricants include mineral oils. However, due to the non-renewable nature of the sources of these mineral base oils, different alternative sustainable base oils are being explored. Among the different sustainable base oils, vegetable oil-based lubricants have attracted immense attention due to their suitable physical and chemical characteristics used for better lubrication. However, these vegetable oil-based lubricants are associated with different drawbacks like prone to self-oxidation, photo-oxidation and thermal oxidations. Therefore, to improve the practical applications of these vegetable oil lubricants, different physical and chemical modifications must be done. Herein, we have discussed one important approach (development of nanolubricants), which is frequently used to improve the lubrication properties of vegetable oil. In the fabrication of nanolubricants, the incorporation of different additives, particularly nanodimensional materials, is being done to improve the range of lubrication and stability toward oxidation of the vegetable oils. Significant developments have been made in nanolubrication technology but still, there is great room for further development and progress. Here, this is worth mentioning that during the development of nanolubricants, proper care should be taken while choosing nanoadditives, as most of the nanodimensional additives could have negative consequences on the environment due to their toxic nature, thus reducing the essence of developing sustainable bio-nanolubricants.

References

1. Anastas, P. and Eghbali, N., Green chemistry: Principles and practice. *Chem. Soc. Rev.*, 39, 301, 2010.
2. Anastas, P. and Kirchhoff, M., Origins, current status, and future challenges of green chemistry. *Acc. Chem. Res.*, 35, 686, 2002.
3. Anastas, P.T. and Zimmerman, J.B., Peer reviewed: Design through the 12 principles of green engineering. *Environ. Sci. Technol.*, 37, 94A, 2003.
4. The commission of the European communities. Establishing ecological criteria and the related assessment and verification requirements for the award of the community eco-label to lubricants. *Official Journal of the European Union*, L 118/126-134, 2005.
5. Regueira, T., Lugo, L., Fandiño, O., Lopez, E.R., Fernández, J., Compressibilities and viscosities of reference and vegetable oils for their use as hydraulic fluids and lubricants. *Green Chem.*, 13, 1293, 2011.
6. Boyde, S., Green lubricants. Environmental benefits and impacts of lubrication. *Green Chem.*, 4, 293, 2002.
7. Sánchez, R., Franco, J.M., Delgado, M.A., Valencia, C., Gallegos, C., Development of new green lubricating grease formulations based on cellulosic derivatives and castor oil. *Green Chem.*, 11, 686, 2009.
8. Mobarak, H., Niza Mohamad E., N., Masjuki, H., The prospects of biolubricants as alternatives in automotive applications. *Renew. Sust. Energ. Rev.*, 33, 34, 2014.

9. Syahrullail, S., Kamitani, S., Shakirin, A., Performance of vegetable oil as lubricant in extreme pressure condition. *Proc. Eng.*, 68, 172, 2013.
10. Kulkarni, M.G., Dalai, A.K., Bakhshi, N.N., Utilization of green seed canola oil for biodiesel production. *J. Chem. Technol. Biotechnol.*, 81, 1886, 2006.
11. Lathi, P.S. and Mattiasson, B., Green approach for the preparation of biodegradable lubricant base stock from epoxidized oil. *Appl. Catal. B*, 69, 207, 2007.
12. Matthew, T.S., Nader, S., Bigyan, A., Lambert, A.D., Influence of fatty acid composition on the tribological performance of two vegetable-based lubricants. *J. Synth. Lubr.*, 24, 101, 2007.
13. Campanella, A., Rustoy, E., Baldessari, A., Baltanas, M.A., Lubricants from chemically modified vegetable oils. *Bioresour. Technol.*, 101, 245, 2010.
14. Battersby, N.S., Pack, S.E., Watkinson, R.J., A correlation between the biodegradability of oil products in the CEC L-33-T-82 and modified sturm tests. *Chemosphere*, 24, 1989, 1992.
15. Meier, M.A., Metzger, J.O., Schubert, U.S., Plant oil renewable resources as green alternatives in polymer science. *Chem. Soc. Rev.*, 36, 1788, 2007.
16. Nagendramma, P. and Kaul, S., Development of ecofriendly/biodegradable lubricants: An overview. *Renew. Sust. Energ. Rev.*, 16, 764, 2012.
17. Lathi, P.S. and Mattiasson, B., Green approach for the preparation of biodegradable lubricant base stock from epoxidized vegetable oil. *Appl. Catal. B.*, 69, 207, 2007.
18. Quinchia, L.A., Delgado, M.A., Valencia, C., Franco, J.M., Gallegos, C., Viscosity modification of high-oleic sunflower oil with polymeric additives for the design of new biolubricant formulations. *Environ. Sci. Technol.*, 43, 2060, 2009.
19. García Zapateiro, L.A., Delgado Canto, M.Á., Franco Gómez, J.M., Valencia Barragán, C., Ruiz Méndez, M., Garcés Mancheño, R., Gallegos, C., Oleins as a source of estolides for biolubricant applications. *Grasas Aceites*, 61, 171, 2010.
20. Rhodes, B.N. and Johnson, D., *Biobased industrial fluids and lubricants*, S.Z. Erhan and J.M. Perez (Eds.), p. 85, AOCS Press, Champaign, Illinois, USA, 2002.
21. Zhou, F., Liang, Y., Liu, W., Ionic liquid lubricants: Designed chemistry for engineering applications. *Chem. Soc. Rev.*, 38, 2590, 2009.
22. Padmaja, K.V., Rao, B.V., Reddy, R.K., Bhaskar, P.S., Singh, A.K., Prasad, R.B., 10-Undecenoic acid-based polyol esters as potential lubricant base stocks. *Ind. Crops Prod.*, 35, 237, 2012.
23. Fox, N.J. and Stachowiak, G.W., Vegetable oil-based lubricants—A review of oxidation. *Tribol. Int.*, 40, 1035, 2007.
24. Murru, C., Badía-Laíno, R., Díaz-García., M.E., Oxidative Stability of Vegetal Oil-Based Lubricants. *ACS Sustainable Chem. Eng.*, 9, 1459, 2021.
25. Wu, Y.Y., Tsui, W.C., Liu, T.C., Experimental analysis of tribological properties of lubricating oils with nanoparticle additives. *Wear*, 262, 819, 2007.

26. Zin, V., Agresti, F., Barison, S., Colla, L., Gondolini, A., Fabrizio, M., The synthesis and effect of copper nanoparticles on the tribological properties of lubricant oils. *IEEE Trans. Nanotechnol.*, 12, 751, 2013.
27. Zhou, J., Yang, J., Zhang, Z., Liu, W., Xue, Q., Study on the structure and tribological properties of surface-modified Cu nanoparticles. *Mater. Res. Bull.*, 34, 1361, 1999.
28. Padgurskas, J., Rukuiža, R., Kreivaitis, R., Asadauskas, S.J., Bražinskienė, D., Tribologic behaviour and suspension stability of iron and copper nanoparticles in rapeseed and mineral oils. *Tribol. - Mater. Surf. Interfaces*, 3, 97, 2009.
29. Choi, Y., Lee, C., Hwang, Y., Park, M., Lee, J., Choi, C., Jung, M., Tribological behavior of copper nanoparticles as additives in oil. *Curr. Appl. Phys.*, 9, e124, 2009.
30. Denape, J., Karama, M., Delbé, K., *Tribological aspects in modern aircraft industry*, Trans Tech Publications, Switzerland, 2015.
31. Kutz, M. (Ed.), *Engineers' Handbook: Materials and Mechanical Design*, 3rd ed., vol. l, John Wiley & Sons, United States, 2016.
32. Fox, N.J. and Stachowiak, G.W., Vegetable oil-based lubricants—A review of oxidation. *Tribol. Int.*, 40, 1035, 2007.
33. Aluyor, E.O., Obahiagbon, K.O., Ori-Jesu, M., Biodegradation of vegetable oils: A review. *Sci. Res. Essays*, 4, 543, 2009.
34. Jayadas, N. and Nair, K.P., Coconut oil as base oil for industrial lubricants - evaluation and modification of thermal, oxidative and low temperature properties. *Tribol. Int.*, 39, 873, 2006.
35. Liaquat, A., Masjuki, H., Kalam, M. *et al.*, Effect of coconut biodiesel blended fuels on engine performance and emission characteristics. *Proc. Eng.*, 56, 583, 2013.
36. Campen, S., Green, J.H., Lamb, G., Atkinson, D., Spikes, H., On the increase in boundary friction with sliding speed. *Tribol. Lett.*, 48, 237, 2012.
37. Weller Jr., D.E. and Perez, J.M., A study of the effect of chemical structure on friction and wear: Part 2–vegetable oils and esters. *Tribol. Lubr. Technol.*, 57, 20, 2001.
38. Bowden, F.P. and Tabor, D., *Friction and lubrication of solids*, The Clarendon Press, Oxford, 1950.
39. Castle, R.C. and Bovington, C.H., The behaviour of friction modifiers under boundary and mixed EHD conditions. *Lubr. Sci.*, 15, 253, 2006.
40. Zisman, W.A., Friction, durability and wettability properties of monomolecular films on solids, in: *Friction and Wear*, R. Davies, (Ed.), Elsevier, Amsterdam, 1959.
41. Warsaw, C.K., Furey, M.J., Ritter, A.L., Molina, G.J., Triboemission as a basic part of the boundary friction regime: A review. *Lubr. Sci.*, 14, 223, 2002.
42. Scrimgeour, C., Chemistry of fatty acids, in: *Bailey's Industrial Oil and Fat Products*, 6th ed., Six Volume Set, F. Shahidi, (Ed.), John Wiley & Sons, Inc., United States, 2005.

43. Technical Committee of the Institute of Shortening and Edible Oils, Inc, *Food Fats and Oils*, 10th ed., Institute of Shortening and Edible Oils, WA, 2016.
44. Shafi, W.K., Raina, A., Irfan Ul Haq, M., Friction and wear characteristics of vegetable oils using nanoparticles for sustainable lubrication. *Tribol. Mater. Surf. Interfaces*, 1, 27, 2018.
45. Zhou, J., Yang, J., Zhang, Z., Liu, W., Xue, Q., Study on the structure and tribological properties of surface-modified Cu nanoparticles. *Mater. Res. Bull.*, 34, 9, 1361, 1999.
46. Farooq, U., Phul, R., Alshehri, S.M., Ahmed, J., Ahmad, T., Electrocatalytic and enhanced photocatalytic applications of sodium niobate nanoparticles developed by citrate precursor route. *Sci. Rep.*, 9, 1, 1, 2019.
47. Farooq, U., Ahmed, J., Alshehri, S.M., Ahmad, T., High-surface-area sodium tantalate nanoparticles with enhanced photocatalytic and electrical properties prepared through polymeric citrate precursor route. *ACS Omega*, 4, 21, 19408, 2019.
48. Farooq, U., Chaudhary, P., Ingole, P.P., Kalam, A., Ahmad, T., Development of cuboidal $KNbO_3$@ α-Fe_2O_3 hybrid nanostructures for improved photocatalytic and photoelectrocatalytic applications. *ACS Omega*, 5, 32, 20491, 2020.
49. Naaz, F., Farooq, U., Khan, M.M., Ahmad, T., Multifunctional efficacy of environmentally benign silver nanospheres for organic transformation, photocatalysis, and water remediation. *ACS Omega*, 5, 40, 26063, 2020.
50. Phul, R., Shrivastava, V., Farooq, U., Sardar, M., Kalam, A., Al-Sehemi, A.G., Ahmad, T., One pot synthesis and surface modification of mesoporous iron oxide nanoparticles. *Nano-Struct. Nano-Objects*, 19, 100343, 2019.
51. Naaz, F., Farooq, U., Ahmad, T., Ceria as an efficient nanocatalyst for organic transformations, in: *Nanocatalysts*, IntechOpen, London, UK, 2019.
52. Farooq, U., Naz, F., Phul, R., Pandit, N.A., Jain, S.K., Ahmad, T., Development of heterostructured ferroelectric $SrZrO_3$/CdS photocatalysts with enhanced surface area and photocatalytic activity. *J. Nanosci. Nanotechnol.*, 20, 6, 3770, 2020.
53. Chen, Y., Zhang, Y., Zhang, S., Yu, L., Zhang, P., Zhang, Z., Preparation of nickel-based nanolubricants via a facile *in situ* one-step route and investigation of their tribological properties. *Tribol. Lett.*, 51, 73, 2013.
54. Wang, A., Chen, L., Xu, F., Yan., Z., *In situ* synthesis of copper nanoparticles within ionic liquid-in-vegetable oil microemulsions and their direct use as high efficient nanolubricants. *RSC Adv.*, 4, 45251, 2014.
55. Jason, Y.J.J., How, H.G., Teoh, Y.H., Sher, F., Chuah, H.G., Teh, J.S., Tribological behaviour of graphene nanoplatelets as additive in pongamia oil. *Coatings*, 11, 732, 2021.

56. Tomizawa, H. and Fischer, T.E., Friction and wear of silicon nitride and silicon carbide in water: Hydrodynamic lubrication at low sliding speed obtained by tribochemical wear. *ASLE Trans.*, 30, 41, 1987.
57. Li, J., Zhang, C., Ma, L., Liu, Y., Luo, J., Superlubricity achieved with mixtures of acids and glycerol. *Langmuir*, 29, 271–275, 2013.
58. Long, Y., Kuwahara, T., De Barros Bouchet, M.I., Ristic, A., Dörr, N., Lubrecht, T., Dupuy, L., Moras, G., Martin, J.M., Moseler, M., *In situ* synthesis of graphene nitride nanolayers on glycerol-lubricated Si3N4 for superlubricity applications. *ACS Appl. Nano Mater.*, 4, 2721, 2021.

Index

Ablation, 96
Additives, 2, 5–10, 18, 20, 33, 34, 38–40, 42–48, 51–59, 63–65, 67, 98, 209–212, 214, 217–224
Advantages of nanolubricants, 2–3
Al with Pd catalysts, 144
Alcohols, 136, 137, 140, 142
Anodization, 173, 176, 179, 203
Application of nanolubricants, 66–67, 70, 72–83
Applications, 173, 197, 199, 200, 203, 206, 207
Auto-oxidation, 238–240

Ball milling, 21, 48–49, 92
Base oil, 34, 40–41, 45
Biodegradable, 209, 210, 217–219, 224
Biolubricants, 154–159
Biomass, 136, 138, 144

Carbon dioxide, 136
Carbon monoxide, 136
Carcinogenic, 136
Characterization, 173, 174, 191, 199, 201, 204–206
Characterization factors, 69–70
Characterization techniques, 70–85
Chemical vapor deposition, 96, 177, 203
Coefficient of friction, 99
Combustion, 185–187, 202
Concept of lubrication, 232–235
 friction, 232–233
 lubrication regimes, 234–235
 wear, 233
Co-precipitation, 188, 206
Corrosion, 136

Deflection imaging, 76
Deformation, 136
Dichalcogenides, 115
DLVO, 26, 28

EDL, 27
Electrochemical deposition, 181, 203
Electrophoretic deposition, 180
Electrospinning, 92
Ester-based biolubricants, 137, 140
Esterification, 137

Friction, 135, 136

Green synthesis, 189, 205
Greenhouse gases, 136

Hardness Rockwell C, 38
Height imaging, 75–76
Heterogeneous catalysts, 136
Homogeneous catalysts, 136, 138, 140, 141, 143
Hydrogenation, 137, 138, 142–144
Hydrothermal, 22–23, 49, 97

Hydrothermal process, 180
Hypodermic, 92

Inert condensation, 21

Lateral force microscopy, 76–77
Lewis-acid catalysts, 138, 139, 144
Lithography, 93
Lubricants, 209–224
Lubricating materials, 154–155
Lubrication, 99

Magnetic force gradient imaging, 77
Mending, 102
Micelles, 215
Microwave-assisted, 187, 202
Mixed ligand complexes, 139
Morphology and topography analysis, 71–85

Nanofluids, 171, 174
Nanolubricants, 103, 171, 173, 194, 197, 199, 205, 209, 212–214, 217, 219, 222–224, 241–250
Nanoparticle additives, 159–167
Nanoparticles, 18, 21–25, 30, 33–34, 47–59, 212–216, 223
Nonvolatile, 20

One-pot hydrogenation–esterification (OHE) process, 143
One-step method, 24
Opto-electronic, 197
Oxide, 214–216, 220, 222–224

Phase imaging, 76
Photodiodes, 198
Photoluminescence, 195, 206
Photo-oxidation of VOs, 240–241
Precious metal-based complexes, 142, 143
Preparation of nanolubricants, 2–4, 67–69
  chemical reduction method, 4
  mechanical milling method, 4
  microemulsion method, 4
  sol–gel method, 4
  sonication method, 4
Pulsed laser ablation technique, 176
Pulsed laser deposition, 95
Pyrolysis, 21–22, 49

RANEY® Ni, 144
Rolling, 29, 31, 35–37, 39–40, 45, 53

Sed rate is Stokes' law, 30
Sensors, 198, 203, 207
Sol–gel, 173–176, 182, 181, 202
Solvothermal, 22–23, 49
Stern Gouy Chapman's model, 27
Surfactants, 20, 25–26, 40, 46, 104
Synthetic, 208–210

The American Petroleum Institute, 41
Thermal degradation, 20
Thermal oxidation of VOs, 241
Thermolysis, 184
Thermophysical properties, 11–13
  density, 12
  flash point, 12
  pour point, 12–13
  surface tension, 11
  thermal conductivity, 12
  viscosity, 11
Torsional resonance microscopy, 77–79
Transesterification, 140, 221, 224
Triacylglycerol, 210, 218

Tribological mechanisms of nanoparticles, 162–167
  ball-bearing effect, 163
  boron-based nanoadditives, 164–167
  mending effect, 163
  polishing effect, 163–164
  protective film formation, 163
  semimetal-based nano-biolubricants, 164
Tribology, 38, 50–53, 56–57, 216, 223
Two-step method, 23–25

Ultrasonic method, 186

Van der Waals forces, 26–27, 35
Vegetable oil, 210–212, 218–221, 224
Vegetable-oil based biolubricants, 234–236
Viscosity, 208–210, 215–220
Viscosity index, 18

Zinc oxide, 173–175, 197–201, 207

Printed in the USA/Agawam, MA
September 16, 2026

910661.061